Tile-Based Geospatial Information Systems

John T. Sample · Elias Ioup

Tile-Based Geospatial Information Systems

Principles and Practices

John T. Sample
Naval Research Laboratory
1005 Balch Blvd.
Stennis Space Center, MS 39529
USA
john.sample@nrlssc.navy.mil

Elias Ioup
Naval Research Laboratory
1005 Balch Blvd.
Stennis Space Center, MS 39529
USA
elias.ioup@nrlssc.navy.mil

ISBN 978-1-4899-9972-6   ISBN 978-1-4419-7631-4 (eBook)
DOI 10.1007/978-1-4419-7631-4
Springer New York Dordrecht Heidelberg London

Printed on acid-free paper

Springer is part of Springer Science+Business Media (www.springer.com)

To Nicole and Oliver
  - John Sample

To Sarah and Georgette
  - Elias Ioup

Preface

Tile-based mapping systems have grown to become the dominant form of mapping system with the rise of Web-based mapping tools. The origin of this book is a desire to collect all our discoveries, techniques, and best practices for creating a tiled-mapping system into one combined volume. The intent of this text is to provide a comprehensive guide to the theory behind creating a tiled-map system as well as a practical guide to create a concrete implementation.

Stennis Space Center, MS   *John Sample*
May 2010   *Elias Ioup*

Acknowledgements

The authors would like to thank the Naval Research Laboratory's Base Program, program element number 0602435N, for sponsoring this research. Additionally, the following people provided technical assistance without which this book would not have been possible: Perry Beason, Frank McCreedy, Norm Schoenhardt, Brett Hode, Bruce Lin, Annie Holladay, Juliette Ioup, and Hillary Mesick.

Contents

Chapter 1
Introduction

This book is intended to provide the reader with a thorough understanding of the purpose and function of tile-based mapping systems. In addition, it is meant to be a technical guide to the development of tile-based mapping systems. Complex issues like tile rendering, storage, and indexing are covered along with map projections, network communication, and client/server applications. Computer code as well as numerous mathematical formulae are included to provide the reader with usable forms of the algorithms presented in this book.

1.1 Background of Web-Based Mapping Applications

The first Web-based mapping applications were introduced in the mid to late 1990's. They included Yahoo! Maps, MapQuest, and Microsoft's TerraServer. These providers offered mapping applications through a Web browser. Their map navigation systems were rudimentary. Some allowed simple map movements by requiring users to click on navigation arrow buttons surrounding the map view. When users clicked on an arrow, the map moved a predetermined amount in the direction clicked. There were also buttons for zooming in and out. Others allowed users to drag and draw boxes on the map to relocate the map view.

All of these systems had several disadvantages, including slow rendering and downloading of map views because the map view was often represented by a single large image file. Each time the map was moved to the left or right; the entire image would be re-rendered and re-sent to the client even though only a portion of the image was new. However, the interfaces were relatively simple and had several advantages to developers. Basic interfaces were well suited to early Web browsers. The map interface could be written entirely in HTML or with very minimal JavaScript. Second, since all navigations were fixed, map servers could cache rendered maps. Other map viewers adopted a map view and navigation style more similar to desktop GIS systems. These systems were more complicated and used browser plugin technology and development platforms like Java or Flash.

J.T. Sample and E. Ioup, *Tile-Based Geospatial Information Systems:*
Principles and Practices, DOI 10.1007/978-1-4419-7631-4_1,
© Springer Science+Business Media, LLC 2010

Google Maps was introduced in 2005 and dramatically changed the way people viewed maps. Instead of clunky and slow map navigation methods, Google Maps provided what has come to be known as a "Slippy Map" type interface. That interface allowed users to quickly move and zoom the map and yet was written entirely in HTML and JavaScript. Soon many more Web mapping applications appeared with a similar style map interface. Eventually slippy map type interfaces appeared in many places including portable computing devices and cell phones.

A key enabling technology behind this new generation of mapping applications was the concept of tile-based mapping. Mapping applications were made responsive by using background maps that had been broken into smaller tiled images. Those tiles were stored, already rendered, on a central server. Because they were already rendered, they could be sent to clients quickly. The tiles were discretely addressed so they could be cached by Internet caching services and by clients' own browsers. The map images were broken into smaller pieces, so when users navigated the map view, only the new parts of the map had to be resent from the server.

1.2 Properties of tile-based mapping systems

Tile-based mapping systems have several core properties which distinguish them from other types of mapping systems. We have defined what we believe to be those core properties, and they are as follows:

1. Map views are based on multiple discrete zoom levels, each corresponding to a fixed map scale.
2. Multiple image tiles are used to virtualize a single map view.
3. Image tiles are accessible using a discrete addressing scheme.
4. Tiled images stored on a server system are sent to the client with minimal processing; as much processing is done ahead of time as is possible.

The following are important but optional properties of tile-based mapping systems.

1. Tile addressing follows a single global projection.
2. Tiles are primarily distributed using a client/server system architecture.
3. Tiles are organized into relatively few, fixed layers.

1.3 Book Organization

This book is organized to take the reader through the logical development of a complete tile-based mapping system with small detours into important topics along the way. Chapter 2 introduces logical tile addressing schemes that any tile system must implement. It discusses some common schemes used by popular Web mapping systems and defines the common tile scheme that will be used throughout this book.

Chapter 3 gives an overview of the challenges and the techniques used to overcome these challenges to develop client software for tile-based mapping systems. An example client application is shown with source code. Chapter 4 provides extensive background into techniques needed to process source data images into tiled images. Chapters 5 and 6 provide a detailed look at techniques for creating sets of tiled images. Chapters 7 and 8 explain how to efficiently store, index, and retrieve tiled images. Several techniques are detailed, implemented, and benchmarked. Chapter 9 shows the reader how to create a Web based server for tiled images. Chapter 10 introduces and explains map projections within the context of tile based mapping. Chapter 11 explains how vector mapping data can be used in a tile-based environment. Finally, Chapters 12 and 13 are detailed case studies of real-world usage of the techniques presented in this book.

Most chapters include computer code listings in Java and Python. Java and Python are two of the most commonly used programming languages for geospatial programming. Short code segments are interspersed with the chapter text, while longer code segments are placed at the end of each chapter. The code sections are intended to provide readers with example implementations of the algorithms explained in the book.

Chapter 2
Logical Tile Schemes

2.1 Introduction

Tile-based mapping systems use a logical tile scheme that maps positions on the Earth to a two-dimensional surface and divides that surface into a series of regularly spaced grids (see Figure 2.1). The logical tile scheme defines the discrete addressing of map tiles, the method for generating multiple zoom levels of tiles, and the translation method between tile addresses and a continuous geospatial coordinate system.

The logical tile scheme is the foundational element of a tile-based mapping system. It is a multi-resolution, regularly spaced grid. Each scheme is typically tied to a single two-dimensional map projection (for more on map projections see Chapter 9). This addressing scheme allows a tiled image to be accessed directly with discrete coordinates. For example, instead of requesting a map image with a bounding rectangle delineated with continuous real numbers like [-100.0, 30.0] to [-80.0, 40.0], a tile can be requested from a grid with level, column, and row addresses delineated with discrete integer values.

The logical tile scheme consists of a mapping between the address of a tile to the geospatial coordinates for the area covered by the tile. In general, there are several ways to develop a logical tile scheme. We could make custom schemes that match the bounds and dimensions of each individual data set, or we could create a single global tile scheme that can be applied to all data sets.

Each method has its benefits. In developing a logical tile scheme, we have to choose a series of image pixel resolutions, one for each level. If we can develop a new scheme for each data set, then we can choose image pixel resolutions that exactly match the resolution of our data set. Using a global common scheme, we have to use the predefined resolutions. Pre-defined resolutions will force us to rescale our source images to match. If we level down, we sacrifice some native resolution, and if we level up, we are using more storage space than is needed. However, if we use a custom scheme for each different data set, we will have interoperability issues in combining the data sets. Also, client software systems will have difficultly using

J.T. Sample and E. Ioup, *Tile-Based Geospatial Information Systems:*
Principles and Practices, DOI 10.1007/978-1-4419-7631-4_2,
© Springer Science+Business Media, LLC 2010

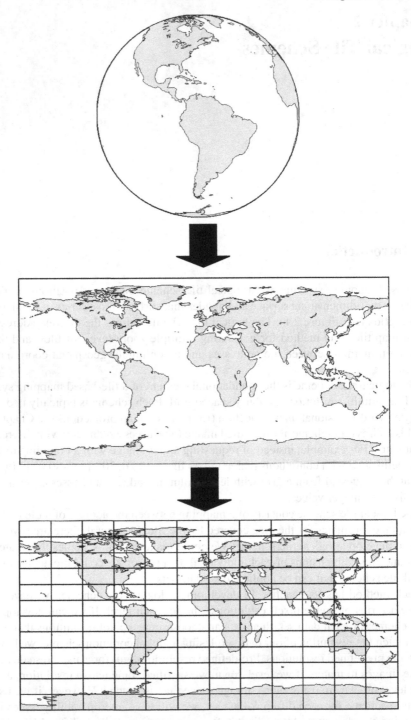

Fig. 2.1 Mapping from a spherical Earth to a two-dimensional surface to a gridded surface.

tiles from different schemes with different resolutions. For our purposes, we choose to use a common global tile scheme that is the same across all data sets. This choice sacrifices flexibility but simplifies system development and use.

2.2 Global Logical Tile Scheme

The global logical tile scheme presented in this book has been developed to be easy to understand and implement. We start with the geodetic projection, which simply portrays the Earth as a rectangle 360 degrees wide and 180 degrees tall. Our base projection has a natural 2-to-1 aspect ratio, and so does our logical tile scheme. At zoom level 1, our tile scheme has 1 row and 2 columns (Figure 2.2).

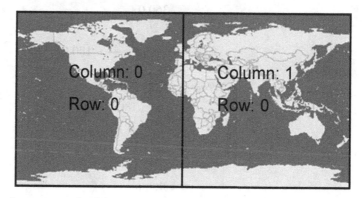

Fig. 2.2 Global tile scheme at zoom level 1.

For each subsequent level, we double the number of rows and columns. Since we have doubled each dimension, each subsequent level has 4 times the number of tiles as the previous level. As we increase zoom levels, each tile is divided into four sub-tiles (Figure 2.3).

We can continue this process and define as many levels as are needed. In practice, 20 levels are sufficient for almost any mapping data available. To simplify the mathematics, we start our indexes at 0 instead of 1. Our scheme can be completely defined mathematically. Equation 2.1 gives the number of columns for a given level i. Equation 2.2 gives the number of rows for a given level i.

$$C_i = 2^i \tag{2.1}$$

$$R_i = 2^{i-1} \tag{2.2}$$

Equations (2.3) through (2.6) relate a tile's address back to a geographic bounding rectangle.

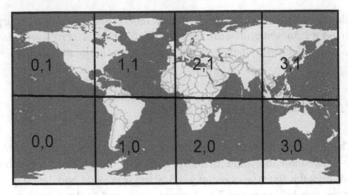

Fig. 2.3 Global tile scheme at zoom level 2.

$$\lambda_{min} = c\frac{360.0}{2^i} - 180.0 \tag{2.3}$$

$$\lambda_{max} = (c+1)\frac{360.0}{2^i} - 180.0 \tag{2.4}$$

$$\phi_{min} = r\frac{180.0}{2^{i-1}} - 90.0 \tag{2.5}$$

$$\phi_{max} = (r+1)\frac{180.0}{2^{i-1}} - 90.0 \tag{2.6}$$

where

$$c \equiv \text{column}$$
$$r \equiv \text{row}$$
$$\lambda \equiv \text{longitude}$$
$$\phi \equiv \text{latitude}$$
$$i \equiv \text{zoom level} \tag{2.7}$$

Chapter 4 discusses choosing the pixel dimensions of tiled images. Once pixel dimensions are chosen, we can compute the resolution of our tiled images in terms of degrees per pixel (DPP). DPP is useful for relating tiled image zoom levels to continuous zoom levels used by many mapping applications. Equation (2.8) is used to calculate degrees per pixel.

$$DPP = \frac{360.0}{2^i}p \tag{2.8}$$

where

$$p \equiv \text{number of pixels per tile}$$
$$i \equiv \text{zoom level}$$

Zoom Level	Number of Columns	Number of Rows	Number of Tiles	Degrees Per Pixel
1	2	1	2	0.3515625000
2	4	2	8	0.1757812500
3	8	4	32	0.0878906250
4	16	8	128	0.0439453125
5	32	16	512	0.0219726563
6	64	32	2048	0.0109863281
7	128	64	8192	0.0054931641
8	256	128	32768	0.0027465820
9	512	256	131072	0.0013732910
10	1024	512	524288	0.0006866455
11	2048	1024	2097152	0.0003433228
12	4096	2048	8388608	0.0001716614
13	8192	4096	33554432	0.0000858307
14	16384	8192	134217728	0.0000429153
15	32768	16384	536870912	0.0000214577
16	65536	32768	2147483648	0.0000107288
17	131072	65536	8589934592	0.0000053644
18	262144	131072	34359738368	0.0000026822
19	524288	262144	137438953472	0.0000013411
20	1048576	524288	549755813888	0.0000006706

Table 2.1 The number of rows, columns, and tiles as well as the degrees per pixel for zoom levels 1 through 20 (assuming 512x512 pixel tiles).

Equations (2.9) and (2.10) show the method for locating the tile that contains a specific geographic coordinate, given a zoom level.

$$c = \left\lfloor (\lambda + 180.0) * \frac{360.0}{2^i} \right\rfloor \tag{2.9}$$

$$r = \left\lfloor (\phi + 90.0) * \frac{180.0}{2^{i-1}} \right\rfloor \tag{2.10}$$

where

$$c \equiv \text{horizontal tile index}$$
$$r \equiv \text{vertical tile index}$$
$$\lambda \equiv \text{longitude}$$
$$\phi \equiv \text{latitude}$$
$$i \equiv \text{zoom level of map view}$$

2.3 Blue Marble Example

To better illustrate the concept, let us begin our first example. We want to define a logical time scheme suited to serving NASA's Blue Marble[1] imagery as tiles. It is satellite derived imagery of the whole earth and provides a good example data set for this book.

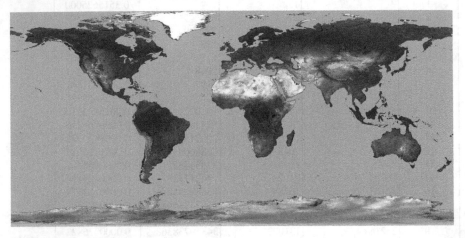

Fig. 2.4 A Blue Marble image.

For our example, we start with a single JPEG image that is 4096 pixels wide and 2048 pixels high. The image covers the entire earth and thus has a bounding rectangle of (-180, -90) to (180,90).

For our example, we are going to use 512x512 pixel tile images. (Chapter 4: Image Processing and Manipulation for GIS will discuss how to choose the proper tile image size for a given application.) Dividing our image width (4096) by our tile width (512) gives us an even 8 tiles across. Likewise, we get an even 4 tiles vertically. From Table 2.1, this is exactly equivalent to zoom level 3. Therefore, we can use zoom level 3 as the base level for our example tile scheme.

In the previous example, our source image matched nicely with our global tile scheme. However, many data sets will not. Suppose we have a single image covering a small geographic area, (-91.5, 30.2) to (-91.4, 30.3), and the image is 1000 by 1000 pixels in size. The image covers a square 0.1 degrees by 0.1 degrees. The resolution of Level 1 is 0.35156. At that resolution, our entire image would only take up 0.28 by 0.28 of a pixel or 7.84% of a pixel. In other words, it would hardly be visible at Level 1. The DPP resolution of the image is 0.1/1000, or 0.0001. This falls between levels 12 and 13 of our global tile scheme. In this case, it might be better to create a custom tile scheme. A method for defining custom schemes is presented in Section 2.5.

---

[1] http://earthobservatory.nasa.gov/Features/BlueMarble

2.4 Mercator-Based Schema

Throughout this book we will focus primarily on tiling systems and data that use the simple Plate Carrée projection, which is also known as the geographic projection. This projection is straightforward to work with and gives us a two-dimensional representation of the earth with a 2-to-1 horizontal to vertical aspect ratio. However, the geographic projection has several shortcomings. At high latitudes, shapes and angles become distorted. To avoid this distortion, many tiling systems use a different base projection for their tiling schemes. The spherical Mercator projection is used by Google Maps, Microsoft Bing Maps, and Yahoo! Maps. Chapter 9 will discuss the details of the Mercator projection. For the purposes of defining a tiling scheme, this projection is significant because it yields a global two-dimensional representation of the earth with a 1 to 1 aspect ratio (see Figure 2.5).

Fig. 2.5 Mercator projection.

Google, Microsoft, and Yahoo! all use a global image similar to what is shown in Figure 2.5 as the top level image in their tiling schemes. Higher resolution zoom levels are generated by dividing each tile into 4 sub-tiles. The only significant difference between these three schemes is in their respective methods for addressing and numbering the tiles. Google Maps uses a simple pair of coordinates to address tiles for a specific zoom level. They set the origin at the top, left of the map. Figure 2.6 shows the addressing for Google Maps at their zoom level 1.

Microsoft's Bing Maps also uses the top-left for its origin but uses a sequential numbering scheme, as shown in Figure 2.7. As the zoom level increases, each tile is divided into 4 sub-tiles. The sub-tiles are sequentially numbered 0 to 3, and that number is concatenated to the number of the parent tile to form the address of the

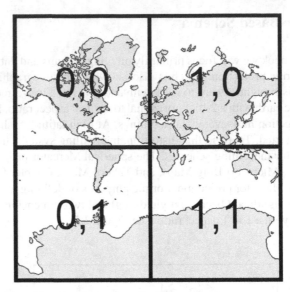

Fig. 2.6 Google Maps tile addressing at zoom level 1.

sub-tiles. Tile 0 is divided into sub-tiles 00, 01, 02, and 03 as shown in Figure 2.8. So, a tile at the 17th zoom level 17 would have 17 digits, one for each zoom level. This numbering scheme makes computing the addresses of sub-tiles trivial. However, relating tile addresses to geographic coordinates, and vice-versa, will require much more computation than the other methods of addressing tiles.

2.5 Variable Start Tile Schemes

NASA World Wind is a freely available virtual globe software system. It provides native support for tiled image sets as map backgrounds on the globe. The default World Wind tile system is very similar to the logical tile scheme we presented in Section 2.1. Like our scheme, World Wind uses the geographic projection. It also uses the bottom-left of the earth as its origin for tile addressing. It differs from our system in that rather than a 2 x 1 tile matrix for its first zoom level, it uses a 10 x 5 matrix as shown in Figure 2.10.

Yahoo! Maps uses a method very similar to Google Maps, except they set their origin tile at the left, middle of the earth, see Figure 2.9.

The World Wind system also allows tiled data sets that do not start at the global level. They use a concept called "Level Zero Tile Size" (LZTS) to define the dimensions of a tile at the lowest resolution level. Then they define a start scale to set the custom tile set's start point. Tiles at the zero level of their default tile scheme are 36.0 degrees by 36.0 degrees, so the LZTS for that scheme is 36.0. The start point

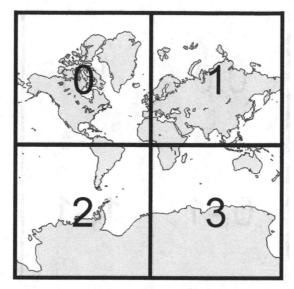

Fig. 2.7 Microsoft Bing Maps tile addressing at zoom level 1.

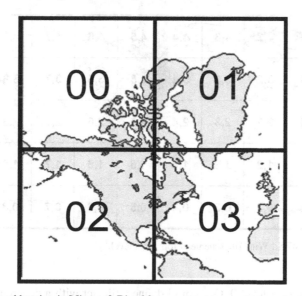

Fig. 2.8 Subtile addressing in Microsoft Bing Maps.

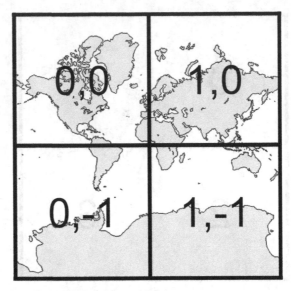

Fig. 2.9 Yahoo! Maps tile addressing at zoom level 1.

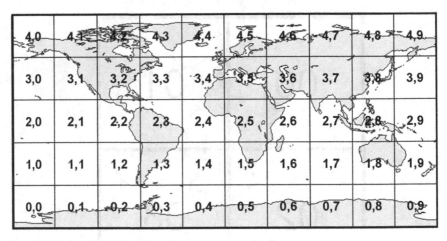

Fig. 2.10 NASA World Wind tile scheme at top zoom level.

for that scheme is at level 0. One could define a tile set with a different LZTS that did not start until a later zoom level, such as 1, 5, or 10. Because of the way World Wind renders the tiled images, custom LZTS values should be a factor of 180.

2.6 Standardized Schema

There are several efforts underway to standardize tile schemes and the way they are communicated. The Web Map Tile Service (WMTS) standard was recently finalized by the Open Geospatial Consortium [1]. It provides a standard but flexible way of defining the capabilities of a tile service and how to interface with it. WMTS does not require the use of one specific tile scheme, resolution set, or projection. Instead, it provides a standard means of defining these properties so clients and servers may be connected together. The WMTS standard does address tiles using matrix coordinates; the top-left tile is addressed as (0,0). However, other properties of the tile scheme are left to the service creator. Multiple different projections are allowed, including the Geodetic and Mercator projections. No restriction is made on which tile scales are made available, only that they be defined using a map scale, meaning the ratio of a distance on the map to a distance on the ground. The map scale is intended only as an identifier for a given zoom level, since it is accurate only near the equator. Tile size may vary over scale, and there may be no relation between the tile matrix dimensions and the scales. Of course, allowing this level of flexibility increases the difficulty of writing a generic client to support a generic WMTS server. To reduce this complexity, the WMTS standard supports a set of well known scale sets that a server may support. By implementing a well known scale set, the server becomes compatible with a wider range of clients. The set of scales in our tile scheme and the set of scales in the Google Maps Mercator tile scheme are both included WMTS well known scale sets. WMTS supports Key-Value-Pair, RESTful, and SOAP request formats for accessing tiles.

Another attempt to create a tile service standard is the Tile Map Service (TMS) specification [2]. The TMS specification is not backed by a standards body but has achieved some level of common usage with a number of servers and clients. It is similar to the WMTS standard in that it allows multiple different tile schemes to be specified. The TMS specification allows the use of arbitrary scales defined by units per pixel. The origin tile may be specified by the server unlike in WMTS where it is always the top-left tile. The tile size may be specified as well. As with WMTS, it supports profiles that specify a map scale and map projection. Both our Geodetic tile scheme and Google Maps Mercator tile scheme are supported profiles. The TMS format supports only a RESTful URL request for tiles.

References

1. Joan Masó, K.P., Julia, N.: OpenGIS Web Map Tile Service Implementation Standard. Open Geospatial Consortium Specification (2010)
2. Ramsey, P.: Tile map service specification. URL http://wiki.osgeo.org/wiki/Tile_Map_Service_Specification

Chapter 3
Tiled Mapping Clients

A tiled mapping client has the responsibility of composing individual map tiles into a unified map display. The map display may allow the user to move around and load in more data, or it may be a static map image whose area is pre-determined by the application. These clients are generally not difficult to create (one of the benefits of tiled mapping). A tiled map client must be able to perform the following tasks:

- Calculate which tiles are necessary to fill the map.
- Fetch the tiles.
- Stitch tiles together on the map.

These three functions are usually performed in sequence as a response to an event in the map client (such as the user moving the map).

3.1 Tile Calculation

The first task of a tiled map client is to calculate which tiles are necessary to fill the map view. The map view is defined by both the geographic area of the map as well as the pixel size of the map. To perform the tile calculations, a simple function is necessary that takes the map view as input and returns a list of tiles as output. Each tile is defined by a tile scale, row, and column.

The actual implementation of such a function is dependent on the way map scale is handled within the client. For a tile client, the simplest method is to allow only a discrete set of map scales. The allowed scales are identical to the set of zoom levels provided natively by the underlying map tiles. Alternatively, the tile client may allow map scales that are not natively available in the tiled map. Usually, such a client will support a continuous set of map scales. Continuous map scales are used in GIS clients because their primary purpose is to create and analyze a wide variety of geospatial data. To support this extensive functionality, users must be able to work on data at any map scale. Supporting continuous map scales means that any combination of geographic area and map size is allowed. As a consequence, the

J.T. Sample and E. Ioup, *Tile-Based Geospatial Information Systems:*
Principles and Practices, DOI 10.1007/978-1-4419-7631-4_3,
© Springer Science+Business Media, LLC 2010

Listing 3.1 Calculation of tile range for the map view.

```
1 numMapViewTilesX= mapWidthPixels / tileWidthPixels
2 numMapViewTilesY = mapHeightPixels / tileHeightPixels
```

function that calculates which tiles are needed is more complex than the function used for discrete map scales.

3.1.1 Discrete Map Scales

The case where clients support only discrete map scales is the simpler one, so that is the best place to begin. As discussed in the previous chapter, the map scale of tiled imagery is specified by zoom level, which is not the traditional distance ratio (e.g., 1:10000m) common on paper maps. Instead, zoom level is simply used to specify the sequence of map scales supported by the tiles. Assuming a standard power of two tile scheme as discussed in the previous chapter, the world will be split into 2^{level} columns and $2^{level-1}$ rows ($level \in \mathbb{N}$). As level increases, so too does map scale. In this case, each increase in level squares the number of pixels used to represent the entire Earth. A map client that uses discrete scales will allow the user to choose only the levels made available by the data.

To calculate which tiles to retrieve, the client need only know the current zoom level, the tile index of the origin of the map view, and the number of tiles required to fill the map view. Notice that the map client need not know the geographic area of the map view. When the map client allows only discrete zoom levels, the state of the map may be stored using only a tile-based coordinate system rather than geographic one. It is also important to note that the following tile calculations assume that the map size is an integer multiple of the tile size. This assumption will be explained further later in the chapter, but it is a result of using common user interface programming techniques.

The origin of the map view is the index of the minimum tile. The minimum tile is the lower left tile when using Cartesian tile coordinates or the upper left tile when using matrix/image (row-major) coordinates. The tile dimensions of the map view are the width and height represented in tiles, i.e., the map width (height) in pixels divided by the tile width (height) in pixels. Listing 3.1 demonstrates how to calculate tile dimensions by dividing the map view size in pixels by the tile size in pixel for each dimension. Listings 3.2 and 3.3 demonstrate the process of increasing and decreasing the zoom level of the map. Both a new tile origin and tile range must be calculated when changing the zoom level.

Full-featured map clients will usually have the functionality to convert a geographic coordinate into tile coordinates as shown in Equations (3.1) (3.2).

Listing 3.2 Calculations for increasing the zoom level. When zoom level is increased we allow truncation when dividing.

```
1   # calculate the new horizontal tile origin index
2     tileCenterX = numCanvasTilesX / 2 + tileOriginX
3     newTileCenterX = tileCenterX * 2
4     tileOriginX = newTileCenterX - numCanvasTilesX / 2
5
6   # calculate the new vertical tile origin index
7     tileCenterY = numCanvasTilesY / 2 + tileOriginY
8     newTileCenterY = tileCenterY * 2
9     tileOriginY = newTileCenterY - numCanvasTilesY / 2
10
11  # calculate the new tile dimensions for the level
12    tileRangeX = tileRangeX * 2
13    tileRangeY = tileRangeY * 2
```

Listing 3.3 Calculations for decreasing the zoom level. When the zoom level is reduced we must round instead of truncating when calculating the tile origin.

```
1   # calculate the new tile dimensions for the level
2     tileRangeX = tileRangeX / 2
3     tileRangeY = tileRangeY / 2
4
5   # calculate the new horizontal tile origin
6     tileCenterX = numCanvasTilesX / 2 + tileOriginX
7     newTileCenterX = tileCenterX / 2
8     tileOriginX = int(round((numCanvasTilesX + tileOriginX) / 2.0)) -
                    numCanvasTilesX
9   if (tileOriginX < 0):
10      tileOriginX = 0
11
12  # calculate the new vertical tile origin
13    tileOriginY = int(round((numCanvasTilesY + tileOriginY) / 2.0)) -
                    numCanvasTilesY
14  if (tileOriginY < 0):
15      tileOriginY = 0
```

$$c = \frac{2^i(\lambda + 180)}{360} \tag{3.1}$$

$$r = \frac{2^{i-1}(\phi + 90)}{180} \tag{3.2}$$

where:

$c \equiv$ horizontal tile coordinate

$r \equiv$ vertical tile coordinate

$\lambda \equiv$ longitude; $-180 \leq \lambda \leq 180$

$\phi \equiv$ latitude; $-90 \leq \phi \leq 90$

$i \equiv$ discrete map scale

3.1.2 Continuous Map Scales

Calculating the tile list for a map client with continuous map scales is more difficult. For a continuous scale client, the map view must be defined by the geographic area of the view and size of the view in pixels. From this definition of map view, it will be necessary to determine which zoom level is best used to populate the view and which specific tiles at that level are in the geographic area.

First, the current map scale must be calculated. The current map scale will still not be represented as a traditional distance ratio since this distance ratio varies over the entire world. Instead, the resolution for the map view, in degrees per pixel, will be used to represent map scale. It should be noted that this works only when the degrees per pixel is constant over the entire Earth at a particular zoom level; for certain map projections, this is not true, and the scale should be represented using the native coordinates of the projection (Chapter 10 has further discussion of map projections).

The degrees per pixel, DPP, may be calculated using the geographic area and size of the map view as shown in Equation (3.3).

$$DPP = \frac{DPP_x + DPP_y}{2} \qquad (3.3)$$

where:

$$DPP_x = \frac{\lambda_1 - \lambda_0}{W}$$

$$DPP_y = \frac{\phi_1 - \lambda_0}{H}$$

$\lambda_1 \equiv$ maximum longitude of map view

$\lambda_0 \equiv$ minimum longitude of map view

$\phi_1 \equiv$ maximum latitude of map view

$\phi_0 \equiv$ minimum latitude of map view

$W \equiv$ width of map view in pixels

$H \equiv$ height of map view in pixels

Each zoom level also has a fixed resolution associated with it. The degrees per pixel for each zoom level may be calculated using the tile geographic bounds and tile size with the above formula.

Once the resolution of the current map view is calculated, the process of determining the best zoom level for tile data may begin. Determining which zoom level to use as the source of tiles has important ramifications on image quality and client performance. If too low a zoom level is chosen, then the image will have an inappropriately low resolution and look pixelated. However, if too high a zoom level is chosen, then the client will be required to fetch too many tiles. For each increase in zoom level, the client must fetch four times the number of tiles to create any given map image. Thus, finding the optimal zoom level for a particular map view is important to the overall performance and quality of the map client. Of course, the optimal zoom level for a given map view is the zoom level with the same image resolution.

In general, the map view will not share an image resolution with any zoom level. Normally, the map view resolution will lie between the resolutions of two zoom levels. Of these two, the zoom level with the higher resolution is the best since it will reduce artifacts due to image scaling. However, a 10% margin of error is used when comparing the map view resolution to the resolution of the lower zoom level. If the map view resolution is within 10% of the lower zoom level resolution then the lower zoom level is used. A 10% resolution reduction is not significant visually and will not impact the resulting map image, whereas the four-fold savings in tile requests will provide significant performance improvements for the map client.

Often, the zoom level identified by the above process may not have the tiles necessary to compose the map view. In this case, an additional search must be conducted for the optimal zoom level tile source. If the upper bounding zoom level is not available, then the lower bounding zoom level should be used, even if the resolution is not within the 10% margin of error. If neither is available, then the next closest zoom levels should be checked, starting with the next highest zoom level. At most, only the next two higher zoom levels should be used. Beyond that the I/O costs of using higher zoom levels are prohibitive and should be avoided. Figure 3.1 shows examples of the processes of calculating the zoom level to use for a map view.

Once the appropriate zoom level is chosen as a tile source, the list of tiles covering the map view must be generated. The calculations for generating the tile list are similar to those used in the discrete map scale cases. First, the minimum tile must be calculated and then the tile range. However, in the continuous map scale case, the map view parameters are not integer multiples of the tile parameters. The minimum tile is calculated using the minimum point on the map view, which may be the lower-left or upper-left point depending on the tile coordinate system in use (we assume lower-left). Equations 3.4 and 3.5 are used to calculate the tile containing a geographic coordinate.

$$c = \lfloor (\lambda + 180.0) * \frac{360.0}{2^i} \rfloor \qquad (3.4)$$

$$r = \lfloor (\phi + 90.0) * \frac{180.0}{2^{i-1}} \rfloor \qquad (3.5)$$

where:

$$c \equiv \text{horizontal tile index}$$
$$r \equiv \text{vertical tile index}$$
$$\lambda \equiv \text{longitude of map view}$$
$$\phi \equiv \text{latitude of map view}$$
$$i \equiv \text{zoom level of map view}$$

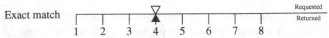

Exact match

(a) The scale of the map view is exactly the same as an available tile zoom level.

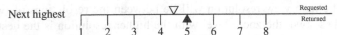

Next highest

(b) The map view scale does not match an available zoom level, so we choose the next highest.

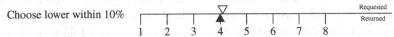

Choose lower within 10%

(c) The map view scale does not match an available zoom level, but it is less than 10% different than the next lowest tile zoom level. As a result we choose the lower tile zoom level.

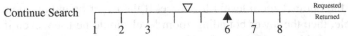

Continue Search

(d) The map view scale lies in between two zoom levels that have no tiles available. In this case, we continue the search and choose the next highest zoom level.

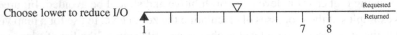

Choose lower to reduce I/O

(e) The map view scale lies in between two zoom levels that have no tiles available. Further surrounding zoom levels are also not available. In order to reduce the number of tiles to retrieve, we choose a lower zoom level, even though higher zoom levels are closer to the map scale.

Fig. 3.1 This figure shows examples of the process for choosing the appropriate zoom level.

The minimum and maximum coordinates of the map view are transformed into minimum and maximum tiles using these formulas. Once the list of tiles required to fill the map view is calculated, the client may proceed to the next major step of retrieving these tiles.

3.2 Tile Retrieval

The tile client must retrieve tiles from where they are stored to use them in the map view. Generally, the tiles are either stored locally or on a network. Sometimes tiles are stored using both mechanisms. Generally, it is a good practice to create an abstract interface for retrieving tiles, so the details of the implementation are separated from the rest of the map client functionality. That way, if the client must change or add an additional retrieval mechanism, the effect on the overall system is minimized.

3.2.1 Local Tile Storage

Local tile storage is the more complicated mechanism for tile retrieval. When tiles on disk are used, the map client must have internal knowledge of the tile storage scheme. In certain cases, the storage scheme is fairly simple. The single file-per-tile scheme is a prime example. Each layer, scale, row, and column combination identifies a single image file that can be easily referenced and exploited in the tile software. Also common is the database tile storage scheme. Here, the map client must be able to connect to the database and properly query for tiles. Database connections from software are trivial to implement these days making this method also relatively simple. More complicated are storage schemes where multiple tiles are stored in a single file. This type of storage system for tiles requires the map client to understand the organization of these files and most likely the indices used to find the tiles within them. Tile retrieval is more difficult in this case, but the performance benefits from such a storage scheme may outweigh the complications. Having a database dependency for a map client is not advisable, given the complexity of installing and managing databases. As discussed in chapter 7 on Tile Storage there are speed and space benefits to storing multiple tiles in a single file.

Generally, local storage of tile data should be limited in overall size. As the amount of tile data increases, so do the demands on the physical system supporting this data. It is usually not desirable to make a map client with large system requirements simply to support the accompanying data. Often, map clients will include one or two map layers with a limited base resolution. These are used as overview maps for the system and only provide a limited number of low zoom levels. Better map layers from external (i.e., network) sources are used for higher resolution data.

3.2.2 Network Tile Retrieval

Retrieval of map tiles from the network is a popular mechanism for map clients to get their data. For Web-based clients, it is a requirement. For desktop clients, it reduces the complexity and size of the software install. While network retrieval of tiles may be accomplished in a number of different ways, most commonly tiles are made available via Hypertext Transfer Protocol (HTTP; the protocol used for the Web). Specifically, each tile is retrieved by performing a GET via HTTP (one of five HTTP functions). By using HTTP GET, each tile is made available by a single URL. When accessing tiles over the network, the client need know nothing about the underlying storage mechanism for the tiles on the server. The server may store tiles in a database, as individual files, or in some custom file scheme, but this

Listing 3.4 Retrieving data from a URL in Python.

```
1  import urllib2
2  urlConnection = urllib2.urlopen('http://host.com/path/tile.jpg')
3  data = urlConnection.read()
```

is not reflected in the URL. Additionally, most programming languages provide libraries, which make retrieving data from a URL using an HTTP GET request a trivial process. Listing 3.4 contains an example of retrieving data from a URL in Python.

There are two common URL styles used in tile retrieval systems. The first encodes the tile parameters in the URL path. This method mirrors the path structure often used when storing tiles as individual files on the file system. An example path encoded URL is `http://host.com/tiles/bluemarble/3/5/2.jpg`. Here bluemarble is the layer name for the tiles, 3 is the zoom level, 5 is the tile column, and 2 is the tile row. The order of these parameters may change, especially the row and the column positions. The second URL style encodes the tile parameters in the URL parameters (the key-value pairs after the ? in the URL). The same map request encoded with URL parameters is `http://host.com/tiles?layer=bluemarble\&level=3\&col=5\&row=2`. Of course, both methods of encoding map requests in a URL have countless variations.

Layer management is another consideration for tile-based map clients to access network stored tiles. The simplest method of layer management is to simply hard-code the list of available layers inside the client. Hard-coding a layer list has the benefit of simplicity. No additional code is necessary to determine the list of layers available. However, hard-coding a layer list can be brittle. Whe the available layers changes, the map client must be updated to support the new layer list. Otherwise, the map client will create an error when a user tries to access a non-existent layer. In many cases the map client may be accessing data maintained by a third party. Without constant vigilance watching for changes in available layers, it is very likely that users will experience data problems.

A more robust alternative to hard-coded layers is to auto-detect the capabilities of each map tile service used by the client. The client may dynamically refresh the available services and layers so the user is always presented with a valid layer list. The result is fewer errors because of service changes or failures. Auto-detecting service capabilities is non-trivial when the tile service has a custom interface. Assuming the service provides a capabilities listing, the map client must have tailored code to parse the capabilities for available layers and supported zoom levels. The Web Mapping Tile Service (WMTS), as discussed in chapter 2, defines a standard capabilities listing format so clients may parse the capabilities of any compliant service.

The map client may choose to manage network errors and performance to improve the user experience. A complete loss of network access prevents the proper operation of a tiled map client, but limited functionality may be maintained by caching commonly used map tiles. The most commonly used tiled data are associated with low scales. Generally, users start with overviews of the entire Earth or a large area of the earth. These views require only a limited number of tiles since they use the lower zoom levels. Network performance may be improved by using tiles from a previous zoom level while a new one is loaded. When a user zooms the map view, the existing tiles may immediately be resized to fill the view. As new tiles come in, they may be placed above the zoomed lower level tiles. Once all tiles for the current zoom level are received, the zoomed tiles may be removed from the map.

3.3 Generating the Map View

After the map client retrieves tiles, it must use them to fill in the map view. The process of assembling tiles into a single map image varies depending on the technologies used to build the map client. However, the underlying process of generating a unified map view is essentially the same.

3.3.1 Discrete Scales Map View

Composing tiles into a single map view is simplest for discrete map scales. The process of determining which tiles to use was discussed above. Once the appropriate tiles are retrieved, they must be combined to form a single map view. The map view may simply be a static image or, more commonly, a portion of the user interface in map client software. Regardless, the algorithm to create the composite view is the same and relatively simple.

For the purpose of this section, we will assume the map client is a program that allows the user to interact with the map. The client must take the retrieved tiles and place them inside the map. The map will be the user interface container that holds the tiles. As stated earlier, the size of the map is an integer multiple of the tile size. The reasoning behind this assumption will be further discussed below, but for now we will hold it to be true. The user interface container used to hold the map varies depending on the programming language used to create the map client. In Java, the container would be a Panel or JPanel object (AWT and Swing respectively). In Python with the Tkinter user interface library, a Canvas object is used to hold images. Other popular programming languages have similar constructs. The following are a list of properties tht must be met by the container for it to function as a map:

- Hold multiple images.
- Absolutely position the images.
- Allow resizing of the container.

Assuming these conditions are met, the container will function appropriately as a map. It should be noted that an image has the above properties and may function as a map.

Placing the image tiles in the container is simple. The horizontal position of a tile is simply the tile index c multiplied by the tile size W. The vertical position must take into account the fact that the container most likely uses matrix coordinates; the upper-left corner is the origin. If the tile scheme uses Cartesian coordinates (the origin is at the lower-left) then a transformation must be made when placing the tiles in the container ($H_{contianer}$ is the container height, r is the vertical tile index, and H_{tile} is the tile height):

$$H_{container} - (r+1)H_{tile} \qquad (3.6)$$

Once the coordinates for each tile are calculated, it may be placed in the container using absolute positioning.

The container holding the tile images is not the same as the map client application window. The application window, which we will call the viewport, will hold the tile container. The viewport is what the user actually sees. The viewport may be smaller or larger than the map container. Separating the size of the viewport from the size of the map container allows the map container to be fixed at an even multiple of the tile size, while the viewport size may vary arbitrarily. The map container will change size as the viewport is resized. The container should have width and height greater than or equal to the width and height of the viewport. At a minimum, round up the viewport width and height to the nearest multiple of tile size to calculate the map container dimensions. Often the map container will be sized to allow an unseen border of tiles one or two tiles deep. The border is used as a tile cache so that when the user moves the map, the tiles will appear on the map without requiring them to be fetched from the tile store. By prefetching unseen tiles, the apparent performance of a network tile store may be significantly improved.

3.3.2 Continuous Scales Map View

When the map client supports continuous scales, the tiled imagery may not be placed directly on the map. Instead, the tiles must be transformed to fit into the current map view. This task may be accomplished by performing the following three steps.

1. Stitch the tiles together into one large image.
2. Cut the large image to match the geographic area of the current map view.
3. Resize the cut image to the current pixel size of the map view.

These three steps are always required; however, some programming languages may provide user interface frameworks that simplify one or more of these steps. For this section we will assume the most basic of built-in functionality.

Stitching the tiles together may be accomplished by pasting them into one larger image. Stitching tiles together in an image is basically the same process as placing

Listing 3.5 Paste tiles into a larger image.

```
1  # get the tile bounds from the geographic bounds
2  minTileX, minTileY, maxTileX, maxTileY = getTileBounds(bounds)
3
4  # retrieve the tile images from the datastore
5  tiles = fetchTiles(minTileX, minTileY, maxTileX, maxTileY)
6
7  # make PIL images
8  tileImages = makeImages(tiles)
9
10 # get the image mode (e.g. 'RGB') and size of the tile
11 mode = tileImage[0][0].mode
12 tileWidth, tileHeight = tileImage[0][0].size
13
14 # calculate the width and height of the large image to paste into
15 largeWidth = tileWidth * (maxTileX - minTileX + 1)
16 largeHeight = tileHeight * (maxTileY - minTileY + 1)
17
18 # make the new image with the correct mode, width, and height
19 largeImage = Image.new(mode, (largeWidth, largeHeight))
20
21 # loop through tiles and paste them into the large image
22 for row in xrange(maxTileY - minTileY + 1):
23     for col in xrange(maxTileX - minTileX + 1):
24         # calculate the location to paste the image (we use Cartesian
25         # tile coordinates)
26         x = col * tileWidth
27         y = largeHeight - ((row+1) * tileHeight)
28         # paste tiles into large image
29         largeImage.paste(tileImage[row][col], (x,y))
```

them in the map container for a discrete scale client. First, make an empty image whose size is the combined width and height of all the retrieved tiles. Each tile should be pasted into the image according to its tile index. As with the discrete zoom level client, the tile indexing affects these calculations. If the tiles are Cartesian indexed (lower-left origin) then the vertical index must be transformed to align with the matrix indexing of images (upper-left origin). Listing 3.5 shows an example of stitching together images with Cartesian tile coordinates.

After the large image is created, it must be cut to match the geographic bounds of the map view. Each corner of the map view has a geographic coordinate. Each coordinate has a pixel location inside the stitched tile image. Those pixel coordinates are then used to cut the large image so its geographic bounds match those of the map view. Listing 3.6 contains code showing the process of cutting the map view.

The final step in creating an image to fill a continuous scale map view is to resize the cut image to have the same pixel size as the map view. First, the resolutions of the map view and the large image are calculated. The scaling factor for the large image is the ratio of the two resolutions. Usually, the resolution is represented as degrees per pixel. Listing 3.7 is an example of resizing the cut image to match the pixel size of the map view.

Listing 3.6 Cut the large image to match the geographic bounds of the map view.

```
1  # assume tileWidth = tileHeight
2  imageDPP = 360.0 / ((2 ** scale) * tileWidth)
3
4  # calculate the pixels for the rectangle to cut
5  leftPixel = int(round((mapViewMinX - imageMinX) / imageDPP))
6  lowerPixel = largeHeight - int(round((mapViewMinY - imageMinY) / imageDPP))
7  rightPixel = int(round((mapViewMaxX - imageMinX) / imageDPP))
8  upperPixel = largeHeight - int(round((mapViewMaxY - imageMinY) / imageDPP))
9
10 # cut out the rectangle
11 cutImage = largeImage.crop((leftPixel, upperPixel, rightPixel, lowerPixel)
```

Listing 3.7 Resizing the stitched together tiles to match the resolution of the map view.

```
1  # assume tileWidth = tileHeight
2  imageDPP = 360.0 / ((2 ** zoomLevel)*tileWidth)
3
4  # the view degrees per pixel may be calculated using the screen size
5  # and geographic bounds
6  viewDPP = getViewDPP()
7
8  scalingFactor = viewDPP / imageDPP
9  newWidth = int(cutWidth * scaling_factor)
10 newHeight = int(cutHeight * scaling_factor)
11 resizedImage = cutImage.resize((newWidth, newHeight))
```

3.4 Example Client

The code in Listing 3.8 contains a working example tile map client. This client
is intended as an example to demonstrate some of the concepts discussed in this
chapter. However, it is extremely simple and should not be considered an example
of a user-ready map client. The example client uses discrete zoom levels and has
limited movement controls. Movement is also limited to one tile at a time. The data
source is a custom tile image server accessible using HTTP over the Internet. This
client should work with a stock Python install along with the Python Image Library
and libjpeg support. A screenshot of the example client is shown in Figure 3.2.

3.5 Survey of Tile Map Clients

A number of popular tile map clients exist and are heavily used by the geospatial
community. Commonly used clients are either Web-based or desktop-based. Most
Web-based clients allow only discrete zoom levels because it simplifies their design.
Performing the image manipulation necessary to support continuous map scales
would be difficult to support in a Web browser as well as overly costly. On the other
hand, Web browsers support discrete zoom levels quite well because of their built-in

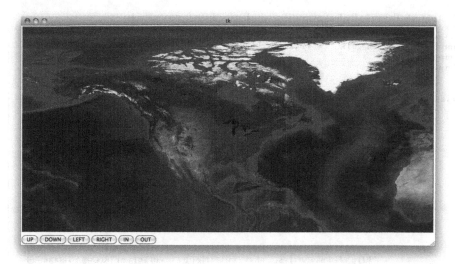

Fig. 3.2 Screenshot of the tile client example.

asynchronous design. When a discrete zoom level map client puts a tile on the map in a browser, it actually gives the image's network location to the browser and tells it to put the image on the map. The browser takes care of fetching the image asynchronously, so that while it is loading, the map client is still functional. Web-based clients are the most commonly used tile map clients. Commercial clients, such as Google Maps and Microsoft Bing Maps, are commonly used by a large non-expert audience to look at maps or get directions. These clients have developer interfaces so that custom data may be added along with the built-in data. However, as these clients are proprietary, they may not themselves be modified by a third party developer. The OpenLayers Web-based map client provides an open source alternative to the proprietary web clients. OpenLayers is a Javascript library that may be used with fewer restrictions than commercial clients.

Two popular desktop map clients are Google Earth and NASA World Wind. Google Earth has significant user penetration and allows developers to add arbitrary tile layers using KML network links. NASA provides World Wind as an open source program which supports full developer customization. Two versions of World Wind exist: a .NET version and a Java version. Both are open source. The .NET version is a full application, whereas the Java version is an SDK, intended for use in building custom map clients. Adding support for custom tile schemes is fairly simple in either.

Listing 3.8 Example Python map client which uses discrete zoom levels.

```
1   import Tkinter
2   import Image, ImageTk     # requires the Python Imaging Library be installed
3   import urllib2
4   import cStringIO
5
6
7   class SimpleNetworkTileSource:
8       '''
9       This is a simple tile data source class to abstract the tile retrieval
            process from the rest of the client. Any class which implements this
            interface (i.e. has the same getTile method and member variables) may
            replace this source in the client.
10      '''
11      def __init__(self):
12          self.layerNames = ['bluemarble']
13          self.tileWidth = 512
14          self.tileHeight = 512
15          self.minLevel = 1
16          self.maxLevel = 8
17
18      def getTile(self, layerName, zoomLevel, colIndex, rowIndex):
19          url = 'http://dmap.nrlssc.navy.mil/tiledb/layerserver?REQ=getimage' + '
                &layer=' + layerName + '&scale=' + \
20              str(zoomLevel) + '&row=' + str(rowIndex) + '&col=' + str(colIndex
                )
21          f = urllib2.urlopen(url)
22          imageData = f.read()
23          return imageData
24
25
26  class MapClient:
27      '''
28      MapClient is an implementation of a simple discrete zoom level map client.
29      '''
30      def __init__(self, parent, tileSource):
31          # Starting zoom level
32          self.level = 2
33
34          # Starting tile origin
35          self.tileOriginX = 0
36          self.tileOriginY = 0
37
38          # The number of tiles in each dimension for this level.
39          self.tileRangeX = 2**self.level
40          self.tileRangeY = 2**(self.level-1)
41
42          # The size of the map view in tiles. We hardcode the size for this
                simple client
43          self.numCanvasTilesX = 2
44          self.numCanvasTilesY = 1
45
46          # The tile source
47          self.tileSource = tileSource
48
49          # This dictionary is used to keep a reference to the tiles displayed in
                the UI so that they are not garbage collected.
50          self.tileImages = {}
51
52          # set up the user interface
53          self.frame = Tkinter.Frame(parent, width=self.tileSource.tileWidth *
                self.numCanvasTilesX, height=self.tileSource.tileHeight * self.
                numCanvasTilesY)
54          self.frame.pack()
55
56          self.canvas = Tkinter.Canvas(self.frame, width=self.getFrameSize()[0],
                height=self.getFrameSize()[1])
```

```
57      self.upButton = Tkinter.Button(self.frame, text="UP", command=self.up)
58      self.downButton = Tkinter.Button(self.frame, text="DOWN", command=self.
           down)
59      self.leftButton = Tkinter.Button(self.frame, text="LEFT", command=self.
           left)
60      self.rightButton = Tkinter.Button(self.frame, text="RIGHT", command=
           self.right)
61      self.inButton = Tkinter.Button(self.frame, text="IN", command=self.
           zoomIn)
62      self.outButton = Tkinter.Button(self.frame, text="OUT", command=self.
           zoomOut)
63
64      self.canvas.pack(side=Tkinter.TOP, fill=Tkinter.BOTH, expand=Tkinter.
           YES)
65      self.canvas.create_rectangle(0,0,self.getMapSize()[0], self.getMapSize
           ()[1], fill='black')
66
67      self.upButton.pack(side=Tkinter.LEFT)
68      self.downButton.pack(side=Tkinter.LEFT)
69      self.leftButton.pack(side=Tkinter.LEFT)
70      self.rightButton.pack(side=Tkinter.LEFT)
71      self.inButton.pack(side=Tkinter.LEFT)
72      self.outButton.pack(side=Tkinter.LEFT)
73
74      # load all the tiles onto the map for the first time
75      self.loadTiles()
76
77  def getFrameSize(self):
78      return (int(self.frame.cget('width')),int(self.frame.cget('height')))
79
80  def getMapSize(self):
81      return (int(self.canvas.cget('width')),int(self.canvas.cget('height')))
82
83  # Below are the controls for moving the map
84  # up, down, left, right, zoom in, and zoom out
85  #
86  # After each is called loadTiles() is called to
87  # refresh the map display with new tiles
88
89  def up(self):
90      if ((self.tileOriginY + self.numCanvasTilesY) < self.tileRangeY):
91          self.tileOriginY += 1
92          self.loadTiles()
93
94  def down(self):
95      if (self.tileOriginY > 0):
96          self.tileOriginY -= 1
97          self.loadTiles()
98
99  def left(self):
100     if (self.tileOriginX > 0):
101         self.tileOriginX -=1
102         self.loadTiles()
103
104 def right(self):
105     if ((self.tileOriginX + self.numCanvasTilesX) < self.tileRangeX):
106         self.tileOriginX += 1
107         self.loadTiles()
108
109 def zoomIn(self):
110     if (self.level < self.tileSource.maxLevel):
111         self.level += 1
112
113         # calculate the new horizontal tile origin index
114         tileCenterX = self.numCanvasTilesX / 2 + self.tileOriginX
115         newTileCenterX = tileCenterX * 2
116         self.tileOriginX = newTileCenterX - self.numCanvasTilesX / 2
```

```
117
118                     # calculate the new vertical tile origin index
119                     tileCenterY = self.numCanvasTilesY / 2 + self.tileOriginY
120                     newTileCenterY = tileCenterY * 2
121                     self.tileOriginY = newTileCenterY - self.numCanvasTilesY / 2
122
123                     # calculate the new tile dimensions for the level
124                     self.tileRangeX = self.tileRangeX * 2
125                     self.tileRangeY = self.tileRangeY * 2
126                     self.loadTiles()
127
128         def zoomOut(self):
129             if (self.level > 2 and self.level > self.tileSource.minLevel):
130                 self.level -= 1
131
132                     # calculate the new tile dimensions for the level
133                     self.tileRangeX = self.tileRangeX / 2
134                     self.tileRangeY = self.tileRangeY / 2
135
136                     # calculate the new horizontal tile origin
137                     tileCenterX = self.numCanvasTilesX / 2 + self.tileOriginX
138                     newTileCenterX = tileCenterX / 2
139                     self.tileOriginX = int(round((self.numCanvasTilesX + self.
                            tileOriginX) / 2.0)) - self.numCanvasTilesX
140                     if (self.tileOriginX < 0):
141                         self.tileOriginX = 0
142
143                     # calculate the new vertical tile origin
144                     self.tileOriginY = int(round((self.numCanvasTilesY + self.
                            tileOriginY) / 2.0)) - self.numCanvasTilesY
145                     if (self.tileOriginY < 0):
146                         self.tileOriginY = 0
147
148                     self.loadTiles()
149
150
151     # calculate which tiles to load based on the tile origin and
152     # the size of the map view (in numCanvasTiles)
153     def calcTiles(self):
154             tileList = []
155             for y in xrange(self.tileOriginY, self.tileOriginY + self.
                        numCanvasTilesY):
156                 for x in xrange(self.tileOriginX, self.tileOriginX + self.
                            numCanvasTilesX):
157                         tileList.append((x,y))
158             return tileList
159
160     # Get the tiles from the tile source and add them to the map view.
161     # This is method is inefficient.  We should really only fetch tiles
162     # not already on the map.  Instead we just refetch everything.
163     def loadTiles(self):
164             tileList = self.calcTiles()
165             self.tileImages.clear()
166             for tileIndex in tileList:
167                 # This is where the tiles are actually fetched.
168                 # A better client would make this asynchronous or place it in
                            another thread
169                 # so that the UI doesn't freeze whenever new tiles are loaded.
170                 data = cStringIO.StringIO(self.tileSource.getTile(self.tileSource.
                            layerNames[0], self.level, tileIndex[0], tileIndex[1]))
171                 im = Image.open(data)
172                 tkimage = ImageTk.PhotoImage(im)
173                 x = (tileIndex[0]-self.tileOriginX) * self.tileSource.tileWidth
174                 y = self.getMapSize()[1] - ((tileIndex[1]-self.tileOriginY+1) *
                            self.tileSource.tileHeight)
175                 self.tileImages[tileIndex] = tkimage
176                 self.canvas.create_image(x, y, anchor=Tkinter.NW, image=tkimage)
```

```
177
178  if __name__ == '__main__':
179      root = Tkinter.Tk()
180      map = MapClient(root, SimpleNetworkTileSource())
181      root.mainloop()
```

Chapter 4
Image Processing and Manipulation

To make source image sets suitable for serving as tiled images, significant image processing is required. This chapter provides a discussion of the image processing techniques necessary to create a tile-based GIS. It discusses algorithms for manipulating, cutting, and scaling different types of images. Several image interpolation algorithms are given with examples and discussion of the relative benefits of each. In addition, this chapter provides guidance for choosing tile image sizes and file formats.

4.1 Basic Image Concepts

A digital image is a computer representation of a two-dimensional image and can be raster or vector based. Raster (or bitmap) digital images use a rectangular grid of picture elements (called pixels) to display the image. Vector images use geometric primitives like points, lines and polygons to represent an image. For the purposes of this book, we are dealing almost exclusively with raster images, which are composed of pixels. Chapter 11 discusses vector data in the context of tiled-mapping.

Each raster image is a grid of pixels, and each pixel represents the color of the image at that point. Typically, the individual pixels in an image are so small that they are not seen separately but blend together to form the image as seen by humans. Consider Figure 4.1; to the left is a picture of a letter A, to the right is that same picture magnified such that the individual pixels are visible.

Pixel values are expressed in units of the image's color space. A color space, or color model, is the abstract model that describes how color components can be represented. RGB (red, green, blue) is a common color model. It provides that the color components for red, green and blue be stored as separate values for each pixel. Combinations of the three values can represent many millions of visible colors. Suppose that we will use values of 0 to 1 to represent each of the components of an RGB pixel. Table 4.1 shows which combinations would create certain common colors.

J.T. Sample and E. Ioup, *Tile-Based Geospatial Information Systems:*
Principles and Practices, DOI 10.1007/978-1-4419-7631-4_4,
© Springer Science+Business Media, LLC 2010

A

Fig. 4.1 Pixelated image.

Red	Green	Blue	Composite Color
1	0	0	Red
0	1	0	Green
0	0	1	Blue
1	1	0	Yellow
0.5	0.5	0.5	Gray
0	0	0	Black
1	1	1	White

Table 4.1 Common colors and their RGB component combination.

Systems that support transparency by means of alpha compositing add a fourth component, ranging from 0 to 1, where 0 indicates that the pixel should be fully transparent, and 1 indicates that the pixel should be fully opaque. This color model is referred to as RGBA or ARGB.

To view and manipulate digital raster images, the RGB components are most often stored as single byte values. In this case, each RGB component is an integer value from 0 to 255 instead of a real value from 0 to 1. The three components will take up 3 bytes (24 bits) or 4 bytes (32 bits) for images with alpha components. It is common to use a 4-byte integer value to store either the RGB or RGBA components.

Raster image pixels are addressable using two-dimensional coordinates with an orthogonal coordinate system over the image. We have used Cartesian coordinates for our mapping coordinate systems, where the center of the coordinates space is $(0,0)$ and horizontal or x coordinate increases as you move to the right and the vertical or y coordinate increases as you move up. Many programming environments reverse the y coordinate such that the origin of an image is at the top-left, and the y coordinate increases as you move down the image. This convention is taken from raster scan based image systems, like cathode ray tube monitors and televisions in which the top-most scanline is the first line displayed for each refresh cycle. The reversal of the y coordinate is an inconvenience that must be considered in all practical applications that relate geospatial data to digital imagery.

Raster images are stored in a variety of file formats defined mostly by their compression algorithms or lack thereof. The most commonly used formats employ compression to reduce the required disk space. Consider an example RGB image that is 1000 by 1000 pixels. To store it uncompressed would require $1000x1000x3 = 3$ megabytes. Its not unreasonable for a good image compression algorithm to obtain a 10 to 1 compression ratio. Thus, the image could be stored in 300 kilobytes.

In general there are two types of compression, lossless and lossy. Lossless algorithms compress the image's storage space without losing any information. Lossy algorithms achieve compression in part by discarding a portion of the image's information. Lossy algorithms seek to be shrewd about what portions of an image's information to discard. Many lossy algorithms can produce a compressed image which discards significant information and yet be visually identical to the original.

The most common lossy image file format is JPEG . JPEG is named after the Joint Photographic Experts Group who created the standard. There are two common lossless file formats: Portable Network Graphic (PNG), and Graphics Interchange Format (GIF). There are several common formats which do not employ compression: Bitmap (BMP), Portable Pixel Map (PPM), Portable Graymap (PGM), and Portable Bitmap (PBM).

4.2 Geospatial Images

Digital images are well suited for storage of geospatial information. This includes aerial and satellite photography, acoustic imagery, and rendered or scanned map graphics. All that is needed to make a digital image a geospatial image is to attach geospatial coordinates to the image in a manner that describes how the image covers the surface of the earth. There are two ways this is commonly done. First, you can provide the bounding rectangle for an image, as in Figure 4.2, or you can provide a single corner coordinate with the resolution of each pixel in each dimension.

Given one or more geospatial images, we can build a tile-based mapping system to distribute the data in those images.

4.2.1 Specialized File Formats

There are several file formats that have been specially adapted for storing geospatial images. MrSID (multi-resolution seamless image database) is a proprietary image

-59.0, 0.0

-99.0,-59.0

Fig. 4.2 Example geospatial image with bounding rectangle defined.

storage format produced by LizardTech. It is specially designed for storage of large geospatial images, most commonly ortho-rectified imagery. MrSID uses a wavelet based compression to store multiple resolutions of the image. This allows for fast access to overview (or thumbnail) sections of the image. It is not uncommon for MrSID images to be generated with many millions of pixels.

JPEG2000 is the next generation file format produced by the Joint Photographic Experts Group. Like MrSID, it is a wavelet based format. JPEG2000 was not specially designed to store geospatial imagery; however, common extensions have been made that allow geospatial information to be attached to the images. JPEG2000 is also well suited to storage of very large images and is a more open format than MrSID.

One of the oldest and most common geospatial image file format is GEOTIFF. GEOTIFF is based on the Tagged Image File Format (TIFF) standard. A GEOTIFF is simply a TIFF file with standard geospatial tags added to it. The TIFF standard is, perhaps, the broadest of any common image file format. It allows many options including alternate compression schemes or no compression at all. It also allows for multi-page images, a variety of color models, and a variety of storage layouts. Fortunately, there are open source software packages for reading and writing TIFF (and GEOTIFF) files that simplify the task of dealing with this complicated image format.

It should be noted that in some cases, geospatial imagery will be stored in files that do not support embedded geospatial coordinates. In those cases, it is customary to provide an accompanying file with the coordinates in it. This is only a convention, not a formal standard. Therefore, the technical details will vary from one implementation to another.

4.3 Image Manipulation

This section will provide background on the image manipulation algorithms needed for the tile creation process, which will be covered in the next chapter. Recall that tiled images are stored in fixed resolutions. It is highly unlikely that a collection of source images will match any single fixed resolution. Since we use multiple resolutions, even if our source images match one resolution, it's impossible for them to match all of our resolutions. Therefore, we are going to have to perform some image scaling.

Image scaling is a type of interpolation. Interpolation is the process of creating new data values within the range of a discrete set of known data values. We will first examine the basic algorithm for scaling and subsetting images. Then we will explain three common interpolation algorithms:

- Nearest Neighbor
- Bilinear
- Bicubic

Each interpolation algorithm has different characteristics with respect to computational performance and output image quality. For the algorithms provided below, we assume that our images have a single color channel. This simplifies the explanation of the techniques. To use the algorithms with three channel color images, the steps are simply repeated for each channel.

The basic component of all of our image scaling algorithms is the same. We will construct a target image, t, and then iterate over the pixels in t, filling them in with data computed from the pixels in our source images. Each image is treated as a two-dimensional array.

The following are some common definitions that will be used in all our image scaling algorithms (see Listing 4.1 and Figures 4.3 and 4.4). The image scaling algorithms will reference a generic interpolation function "interpolate" (Listing 4.2). The first parameters are the details of the source image. "tx" and "ty" are the map coordinates of a pixel that is to be interpolated from the source data. The image scaling algorithms also reference a common function "geolocate" that calculates the geographical coordinate of the center of a pixel (Listing 4.3). Because images are stored in scanline order, the y coordinates have to be flipped. The variable `adj_j` is created to do this.

In the first scaling algorithm (Listing 4.4 and Figure 4.5), we make a simplification assumption that the source image and the target image have the same map coor-

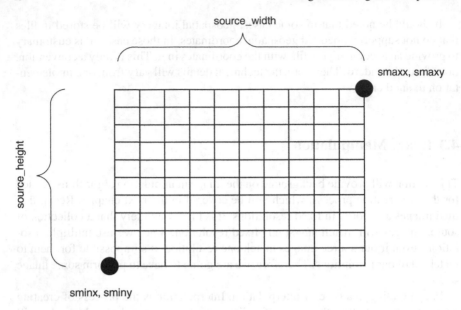

Fig. 4.3 Source image parameters: sminx, sminy, smaxx, smaxy, source_width, and source_height.

Target Image

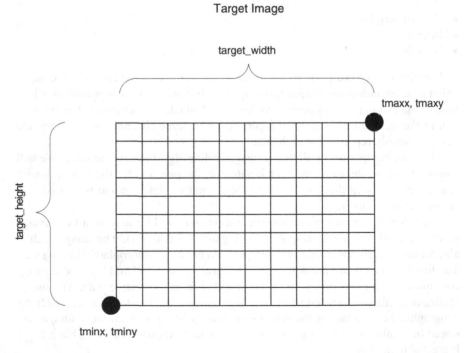

Fig. 4.4 Target image and parameters: tminx, tminy, tmaxx, tmaxy, target_width and target_height.

Listing 4.1 Definitions of variables in code examples.

```
1  integer s[][]:   Source Image, a 2-d array
2  integer source_width: width of source image
3  integer source_height: height of source image
4  real sminx: minimum horizontal map coordinate of source image
5  real sminy: minimum vertical map coordinate of source image
6  real smaxx: maximum horizontal map coordinate of source image
7  real smaxy: maximum vertical map coordinate of source image
8
9  integer t[][]:   Target Image, a 2-d array
10 integer target_width: width of target image
11 integer target_height: height of target image
12 real tminx: minimum horizontal map coordinate of target image
13 real tminy: minimum vertical map coordinate of target image
14 real tmaxx: maximum horizontal map coordinate of target image
15 real tmaxy: maximum vertical map coordinate of target image
```

Listing 4.2 Definition of abstract function interpolate that will be implemented by specific algorithms.

```
1  function integer interpolate(s, sminx, sminy, smaxx, smaxy, source_width,
         source_height, tx, ty)
```

Listing 4.3 Compute the geographic coordinates of the center of a pixel.

```
1  function real,real geolocate(real minx,miny,maxx,maxy, integer i,j,width,height
       )
2      comment:
3          minx,miny,maxx,maxy are the geographical coordinates of the image
4          width and height are the dimensions of the image
5          i and j are the pixel coordinates to be converted to geographic
              coordinates
6
7      real pixel_width= (maxx-minx) / width
8      real pixel_height= (maxy-miny) / height
9
10     real x =(i + 0.5) * pixel_width  + minx
11
12     int adj_j = height - j - 1
13
14     real y =(adj_j + 0.5) * pixel_height + miny
15
16     comment: we offset by 0.5 the indexes, to get the center of the pixel
17
18     return x,y
```

dinates but different dimensions. So, sminx = tminx, miny = tminy, smaxx = tmaxx, and smaxy = ymaxy, but source_width ≠ target_width and source_height ≠ target_height.

Care should be taken in determining whether to iterate in row-major or column-major order. This is a practical consideration that must be made in the context of specific programming environments. Java, C, and many others store array data in row-major format. Iterating in this fashion can potentially greatly improve the performance of the algorithm due to the principle of Locality of Reference. In the con-

Listing 4.4 Simple Image Scaling: source and target images have the same geographic coordinates but different sizes.

```
1  for j in xrange(target_height):
2      for i in xrange(target_width):
3          (tx, ty) = geolocate(tminx, tminy, tmaxx, tmaxy, i, j, target_width,
                  target_height)
4          val = interpolate(s, sminx, sminy, smaxx, smaxy, source_width,
                  source_height, tx, ty)
5          t[j][i] = val
```

Source Image

smaxx, smaxy

sminx, sminy

Target Image

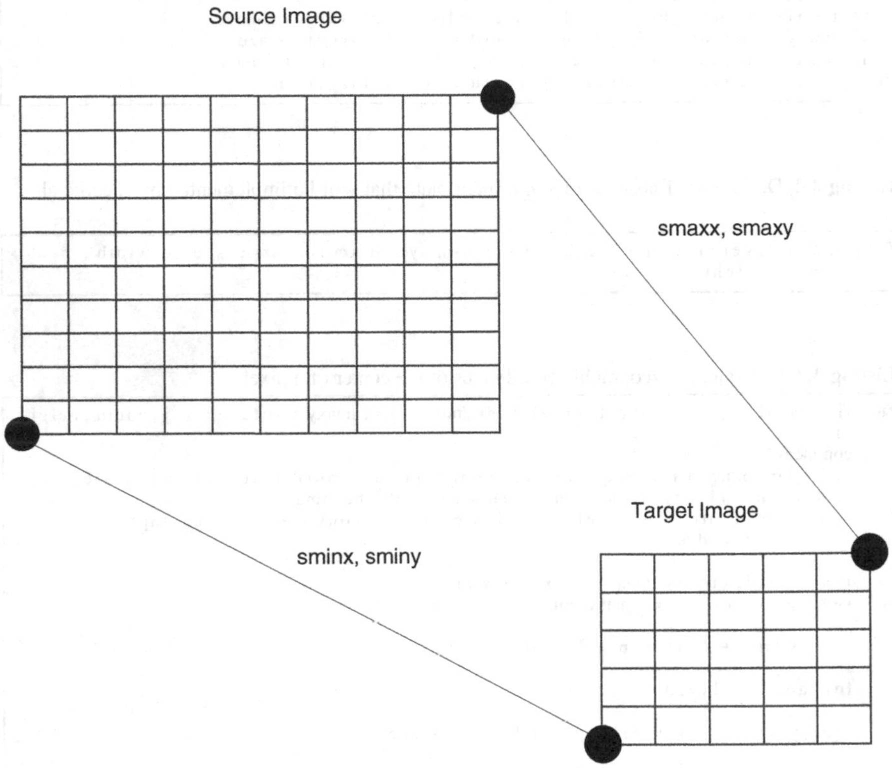

Fig. 4.5 In Scaling Algorithm 1, the source and target images share coordinates.

text of digital image manipulation, it means we should access pixels in roughly the order they are stored in the computer's memory. This reduces the number of times the operating system has to pull new memory pages into the cache [1]. Our second image scaling algorithm (Listing 4.5 and Figure 4.6) is a more general version of algorithm 1. In it, t is a scaled subsection of s. This algorithm is suitable as a basis for almost any rescaling and subsetting task.

Next, we will define our interpolation algorithms. Each of following interpolation algorithms implements the generic "interpolate" function defined earlier. In general, interpolation solves the problem shown in Figure 4.7. That is, we want to get the

Listing 4.5 Target image is a scaled subsection of source image.

```
1  for  j = 0 to target_height − 1,
2      for  i = 0 to target_width − 1,
3              real tx;{tx is the target pixel's x coordinate}
4              real ty;{ty is the target pixel's y coordinate}
5              tx,ty = geolocate(tminx,tminy,tmaxx,
6                          tmaxy,i,j,target_width,target_height)
7
8              integer pixel_val= interpolate(s, sminx, sminy, smaxx, smaxy,
                       source_width, source_height, tx, ty);
9
10             t[j][i] = pixel_val;
11      end if
12 end if
```

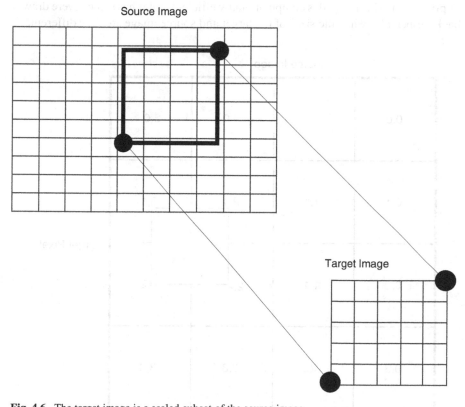

Fig. 4.6 The target image is a scaled subset of the source image.

value for a target pixel that does not correlate exactly to a source pixel. In this case, the target pixel overlaps pixels (2,1) , (3,1), (2,2), and (3,2).

4.3.1 Interpolation 1: Nearest Neighbor

Nearest neighbor is the simplest of all interpolation algorithms. It uses the pixel value from the source image that is the closest spatially to the target pixel's location. Following the graphic in Figure 4.7, we can visually determine that pixel (3,1) is the "closest" to the center of the target pixel. In this case, for nearest neighbor interpolation, the resulting value of the target pixel would simply be the exact value of pixel (3,1). This method is computationally efficient, but it has some severe drawbacks, especially when the sizes of the target and source image are very different.

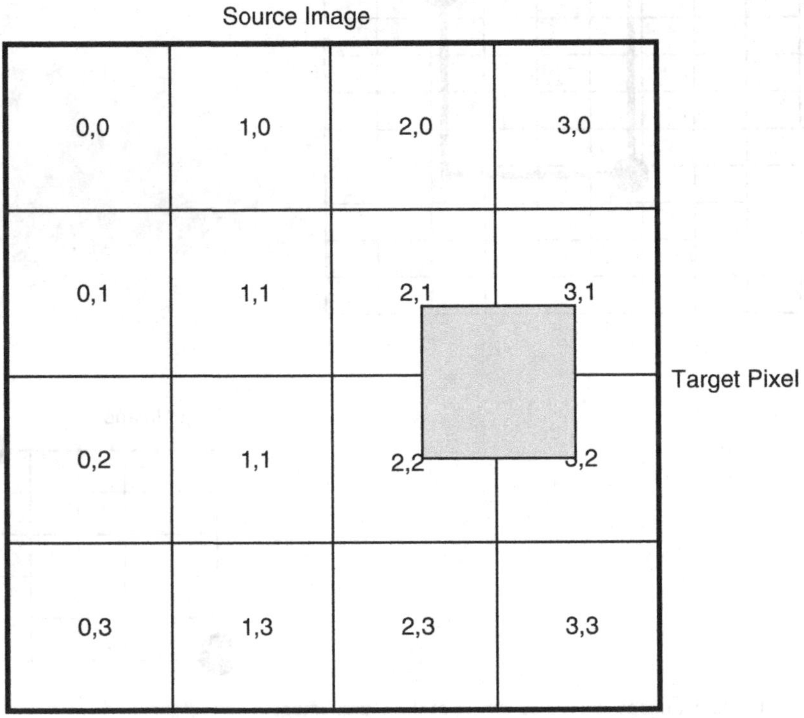

Fig. 4.7 Nearest neighbor interpolation uses only the closest pixel (3,1) to determine the target value.

Listing 4.6 shows the algorithm for nearest neighbor interpolation. The real work in this function is done by the "round" function, which simply rounds a real value to the closest integer.

Listing 4.6 Nearest neighbor interpolation.

```
1  def nearest_neighbor(s, sminx, sminy, smaxx, smaxy, source_width, source_height
       , tx, ty):
2      i = round((tx - sminx) / (smaxx - sminx) * source_width)
3      j = source_height - 1 - round((ty - sminy) / (smaxy - sminy) *
           source_height)
4      return s[j][i]
```

Listing 4.7 Bilinear interpolation.

```
1  from math import *
2
3  def bilinear(s, sminx, sminy, smaxx, smaxy, source_width, source_height, tx, ty):
4      temp_x = (tx - sminx) / (smaxx - sminx) * source_width
5      temp_y = source_height - 1 - ((ty - sminy) / (smaxy - sminy) *
           source_height)
6
7      i = floor(temp_x)
8      j = floor(temp_y)
9      weight_x = temp_x - i
10     weight_y = temp_y - j
11     val_00 = s[j][i]
12     val_01 = s[j][i+1]
13     val_10 = s[j+1][i]
14     val_11 = s[j+1][i+1]
15
16     pixel_val = (1 - weight_x) * (1 - weight_y) * val_00 + weight_x * (1 -
           weight_y) * val_01 + (1 - weight_x) * weight_y * val_10 + weight_x *
           weight_y * val_11
17
18     return pixel_val
```

4.3.2 Interpolation 2: Bilinear

Bilinear interpolation is a little more complicated; it creates a weighted average of the 4 pixels which surround the center of the target pixel (Listing 4.7). Recall Figure 4.6; the bilinear interpolation would use pixels (2,1) , (3,1), (2,2), and (3,2).

Figure 4.8 illustrates the computations in the bilinear algorithm. The arrowed lines go from the center of the source pixels to the center of the target pixel. The length of each line, in ratio to the sum of the lengths, forms the complement of the weight given to the data from the pixel in which the line originates. It forms the complement because we want pixels with greater length to have less impact on the final result. They are "further away" from the target pixel.

Let's consider a variation on our bilinear algorithm. Suppose that our target pixel covers a large area in our source image, as in Figure 4.9. In this case, the bilinear algorithm would only use pixel data from pixels (1,1), (2,1), (1,2), and (2,2). Data from the other pixels would be disregarded. There are several solutions to this problem. The easiest is to perform multiple interpolation steps. Divide the target pixel into four (or more) sub-pixels and then perform a bilinear interpolation for each sub-pixel. When that is complete, compute the final target pixel value by bi-

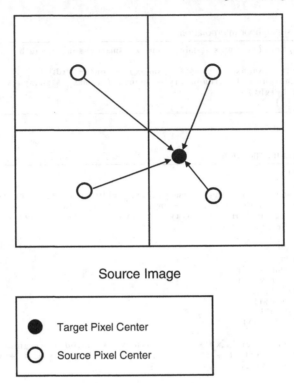

Source Image

Target Pixel Center

Source Pixel Center

Fig. 4.8 Bilinear interpolation uses the four surrounding pixels to compute the target value.

linear interpolation over the four sub-pixels. This type of multi-step (also called multi-resolution) interpolation is the best way to handle image scaling that shrinks an image by a significant amount (Figure 4.10).

4.3.3 Interpolation 3: Bicubic

Bicubic interpolation is the most complicated of our interpolation algorithms. Where the bilinear interpolation considered the linear relationship of the 4 pixels surrounding our target point, the bicubic algorithm computes a weighted average of the 16 surrounding pixels. Figure 4.11 shows the target pixel with 16 surrounding pixels. Even though the outer 12 pixels do not overlap the target pixel, they are used for computing the surrounding gradients (or derivatives) of the pixels that do overlap the target pixel. This does not necessarily produce a more accurate interpolation, but it does guarantee smoothness in the output image.

The one-dimensional cubic equation is as follows:

$$f(x) = a_0 x^3 + a_1 x^2 + a_2 x + a_3$$

Source Image

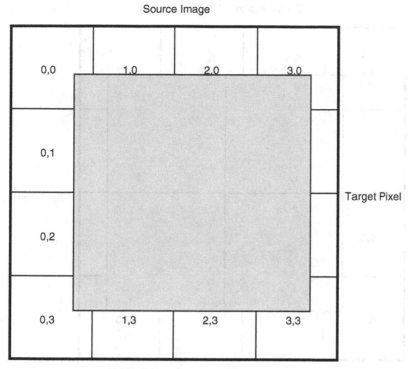

Fig. 4.9 Bilinear interpolation where the target pixel covers a large pixel area of the source image.

There are four coefficients: a_0, a_1, a_2, and a_3. The two-dimensional cubic equation, $f(x,y)$ has 16 coefficients, a_{00} through a_{33}. There are several ways to compute the 16 coefficients using the 16 pixel values surrounding the target pixel. Most involve approximating the derivatives and partial derivatives to develop a set of linear equations and then solving the linear equations. The full explanation of that process is beyond the scope of this chapter. We suggest the references *Numerical Recipes in C* and "Cubic convolution interpolation for digital image processing" for more information [3, 2].

Since each interpolation algorithm has different performance characteristics, we will examine the results with real images. Figure 4.12 is an image of a fish.[1] If we scale a small section of the fish's scales to 400% (or 4 times magnification) in each dimension, we get the images shown in Figure 4.13, Figure 4.14, and Figure 4.15 for nearsest neighbor, bilinear, and bicubic interpolations, respectively. In this example, only the bicubic interpolation yields a satisfactory result.

We will also consider an example using a rendered map graphic. Figure 4.16 is a map of a portion of the city of New Orleans from OpenStreetMap.[2] We will use

---

[1] Fish images courtesy of Robert Owens, Slidell, Louisiana.

[2] OpenStreetMap images used from www.openstreetmap.org.

Source Image

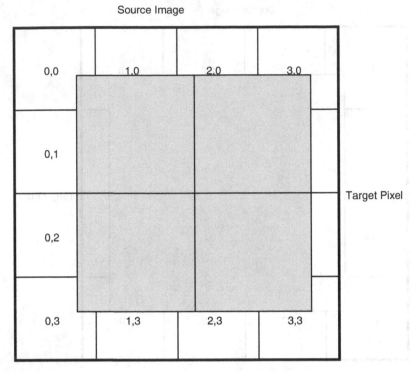

Target Pixel

Fig. 4.10 Bilinear interpolation can be performed in multiple steps to compute target pixels that cover many source pixels.

our interpolation algorithms to scale a sub-section of that image. We have chosen a subsection with lots of lines and text. These are typical features in map images. If we scale a small section of the image to 400% (or 4 times magnification) in each dimension, we get the images shown in Figure 4.17, Figure 4.18, and Figure 4.19 for nearsest neighbor, bilinear, and bicubic interpolations respectively. Once again, only the bicubic provides a satisfactory result. Figure 4.20 shows a section of the image with text highly magnified by bicubic interpolation. Figure 4.21 shows a section of the image with text highly magnified by bilinear interpolation. The bicubic interpolation performs much better with text features.

Text features are especially sensitive to interpolation. Even though the bicubic interpolation imposes a significant performance penalty, it is probably worth the cost in most cases.

Listing 4.9 shows implementations of the nearest neighbor and bilinear interpolation algorithms for RGB images. The classes BoundingBox and Point2DDouble are simply wrapper classes for multiple coordinates. BufferedImage is the Java built-in class for manipulating image data. Many programming environments provide built-in tools for scaling and subsetting images. This changes our algorithms slightly. Instead of performing pixel-by-pixel calculations, we compute a single set of trans-

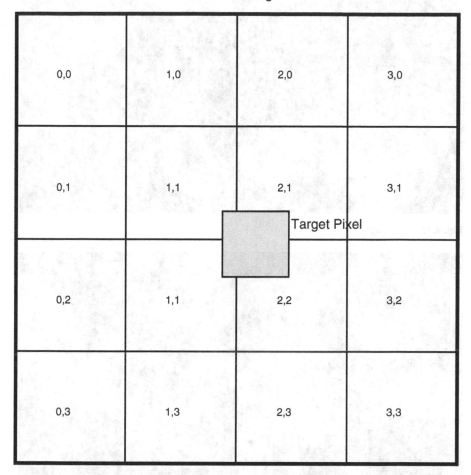

Fig. 4.11 Bicubic interpolation uses the 16 surrounding pixels to compute the target pixel value.

formation parameters and pass those to the built-in image manipulation routines. Listings 4.10 and 4.11 show how to use those built-in routines in Java and Python.

Practical experience has shown that bilinear interpolation takes approximately 150% the time as nearest neighbor, and bicubic interpolation takes approximately 200% the time as nearest neighbor.

The astute reader will notice that we have used bilinear interpolation throughout our discussion as the means of calculating the geographical coordinates. The supplied algorithm "geolocate" uses bilinear interpolation to map between geographic and pixel coordinates. So why is it good enough to use bilinear for calculating geographic coordinates but not good enough for calculating the actual pixel values? The mapping from geographical coordinates to pixel space and back is, by definition, a

Fig. 4.12 Fish Image

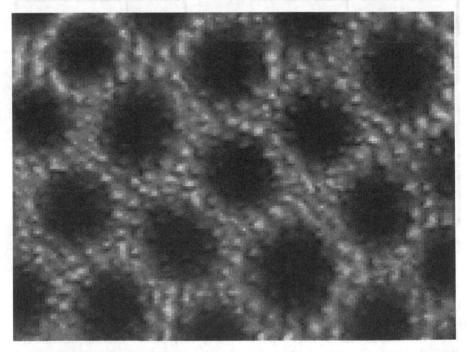

Fig. 4.13 Fish Scale image magnified with nearest neighbor interpolation.

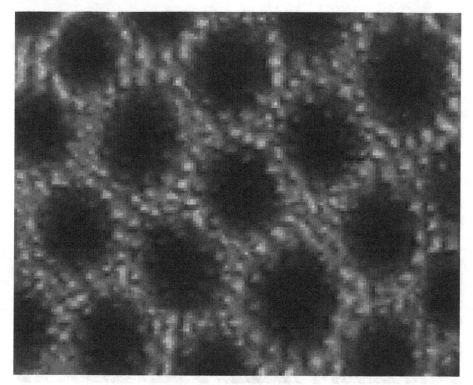

Fig. 4.14 Fish Scale Image magnified with bilinear interpolation.

linear function. We can compute the exact transformation from geographic coordinates to pixel coordinates. However, the contents of the image, the actual color values of pixels, are highly non-linear. Whether the image contains aerial imagery or a rendered map graphic, there is very little linearity, either locally or globally, between the actual values of the pixels.

4.4 Choosing Image Formats for Tiles

Any tile-based mapping system must use image file formats for storage and transmission of image tiles. There are hundreds of file formats that can be used. Some offer very sophisticated compression schemes, and others focus on simplicity and compatibility. We want to choose image formats that can be encoded and decoded quickly, offer good compression performance, and, most importantly, are supported natively by common web browsers.

It is possible that we will use one format to store images and another to transmit them. In general, we want to reduce image processing and manipulation tasks that

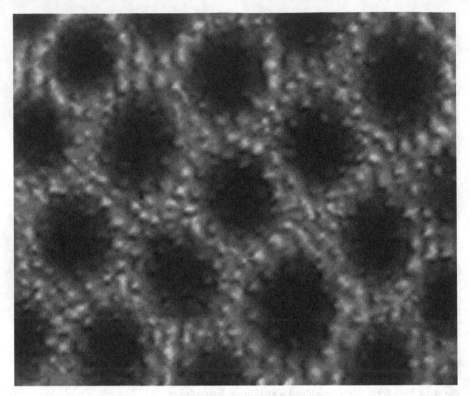

Fig. 4.15 Fish Scale Image magnified with bicubic interpolation.

are required for each client access. Our goal is to use the same format for storage and retrieval.

Table 4.2 lists several popular browsers and their supported image formats. Native browser support is critical. Browser-based (HTML/JavaScript) map clients, like OpenLayers and Google Maps, achieve their quick performance and appealing look by using the native capabilities of the browser to display and manipulate images. If we adopt formats that are not well supported by the majority of Web browsers, we have needlessly crippled our system's performance. From the table, we can see that JPEG, GIF, BMP, and PNG are commonly supported. Table 4.3 shows the features of each format.

Browser	JPEG	JPEG2000	GIF	TIFF	BMP	PNG
Internet Explorer	Yes	No	Yes	No	Yes	Yes
Firefox	Yes	No	Yes	No	Yes	Yes
Google Chrome	Yes	No	Yes	No	Yes	Yes
Safari	Yes	Yes	Yes	Yes	Yes	Yes
Opera	Yes	No	Yes	No	Yes	Yes

Table 4.2 Browser support for different image compression types.

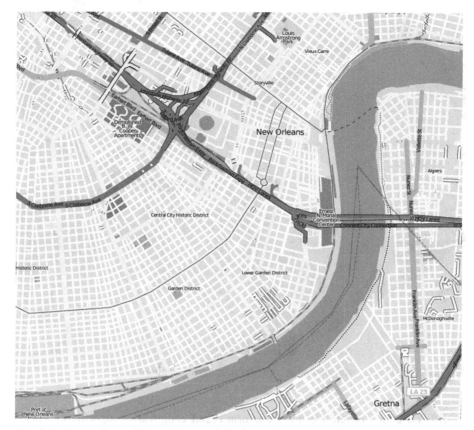

Fig. 4.16 Rendered map of New Orleans.

Format	Compression	Colors Supported	Transparency Supported
JPEG	Lossy	24 bit RGB	No
GIF	Lossless	8 bit Indexed	Yes
BMP	Uncompressed	24 bit RGB	Yes
PNG	Lossless	48 bit RGB	Yes

Table 4.3 Details of different compression types.

We can eliminate BMP from consideration since it is not compressed. Also, we can eliminate GIF because it does not lend itself to full 24 bit color. This leaves us with PNG and JPEG. PNG provides lossless compression and support for transparency while JPEG provides lossy compression.

PNG uses the DEFLATE lossless compression algorithm. PNG can achieve superior compression with images that have few unique colors, repeated pixel patterns, and long sequences of the same pixel value. As such, it is quite suitable for storing rendered figures and maps that typically have limited color palettes.

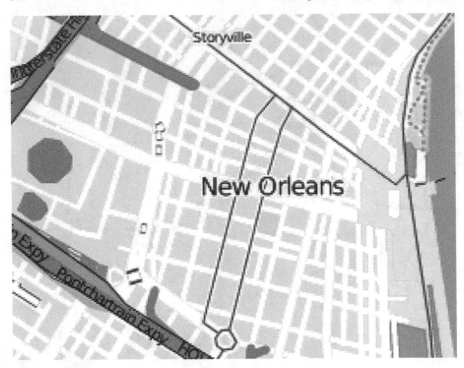

Fig. 4.17 New Orleans map subsection with nearest neighbor interpolation.

JPEG uses a Discrete Cosine Transform based compression algorithm. It performs well with images that have lots of colors, some noise, and softer transitions typically found in photography.

As the basis for a comparison of PNG and JPEG performance characteristic, recall our fish image, Figure 4.12. The color version of this image has 315,559 colors and 2088 by 1128 pixels. Stored as a JPEG, the file is 147kb. Stored as a PNG, the file is 2.67mb. That is a ratio of around 18 to 1. That means if we use PNG storage for our tiles we will need 18 times the storage space, and our users will have to wait 18 times longer for the images to download. The color version of our rendered map graphic of New Orleans, Figure 4.16, has 2372 colors and 780 by 714 pixels. The PNG version is 321kb, and the JPEG version is 113kb. This is a much more reasonable 3 to 1 ratio.

While not visible at the default scales, compression artifacts are visible where there are sharp color boundaries in the image. Figure 4.22 is a test image with some text saved as a JPEG with the default quality settings. Compression artifacts are visible at the text boundaries.

Based on these considerations, we provide the following guidance.

- Use JPEG images when dealing with aerial or satellite photography, images with lots of colors, or when storage space is a critical issue.

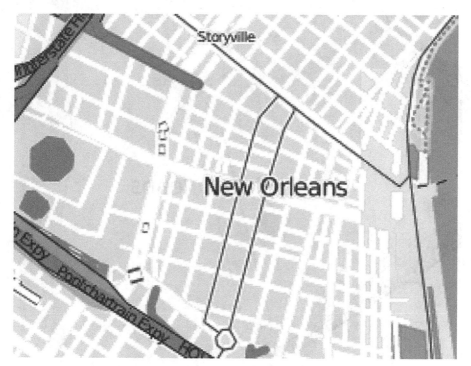

Fig. 4.18 New Orleans map subsection with bilinear interpolation.

- Use PNG images when transparency is required or when quality of reproduction for rendered map graphics is critical.

Both PNG and JPEG include, within the first few bytes of the file format, a unique identifier that allows image reading software to know the format of the file. This self identification property simplifies tile storage, since the tile storage system does not need to store the file type that was used to store a tile.

It is perfectly reasonable to use both formats together in the same tile system. For example, a tiled map layer that has data for only a small portion of the earth would use a transparency enabled format for the low-resolution scales so that map users could see the covered areas in the conjunction with other background layers. It would then switch to JPEG for the high-resolution images that have larger storage requirements.

The reader may ask why we have not chosen to use one the common geospatial image formats for storing our tiles. The answer is simple. First, tiles have their geospatial coordinates embedded in their tile address. Recall from the discussion in Chapter 2 on logical tile schemes that a tile scheme provides for conversion from a tile's address to its map coordinates and back. Secondly, and most importantly, geospatial image formats are not commonly supported by Web browsers.

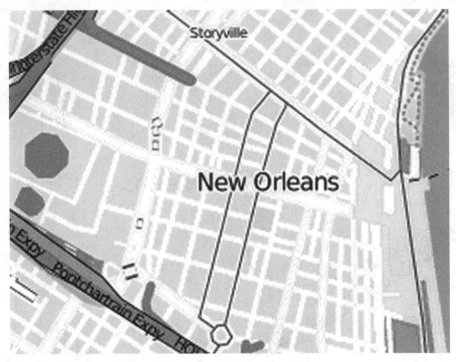

Fig. 4.19 New Orleans map subsection with bicubic interpolation.

Fig. 4.20 Close up text with bicubic interpolation.

Fig. 4.21 Close up text with bilinear interpolation.

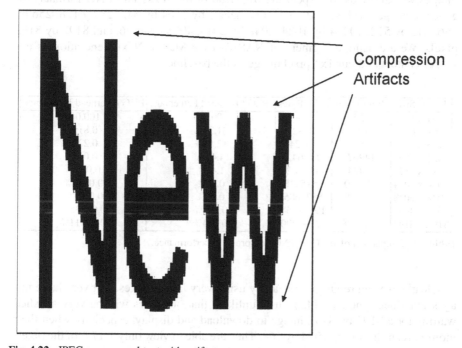

Compression
Artifacts

Fig. 4.22 JPEG compressed text with artifacts.

4.5 Choosing Tile Sizes

The choice of tile image dimensions is one of the most important decisions to be made in the design of a tile-based mapping system. Tile images can be any size, and they can vary from scale to scale. They can also vary across the same scale, or they can be random in size. However, there are efficiencies with making tiles uniform in size across each and every scale. Also, there are efficiencies from choosing tiles that have the same horizontal and vertical dimensions. Furthermore, tile sizes that are powers of two yield simpler mathematics throughout the process.

There are several approaches to determining the optimal tile size. First we should consider the impact of using multiple images to virtualize a single map view. Each image comes with a certain amount of overhead. There are several types of overhead involved that include the overhead of multiple seeks and reads from the computer's file system, uneven utilization of the file system's native block size, and the header and other overhead storage space within each image file.

Let us consider the constraints of current image formats. We have limited ourselves to image formats that are readily usable by most Web browsers: JPEG and PNG. Any encoded image is going to use space for overhead, i.e. space not directly used to store pixels. This is header information and image metadata. Some example images will allow us to inspect the overhead of the JPEG and PNG formats. We generate images with scaled content of sizes 1 by 1, 64 by 64, 128 by 128, 256 by 256, 512 by 512, 10214 by 1024, 2048 by 2048, 4096 by 4096, and 8192 by 8192 pixels. We are using a segment of NASA's Blue Marble Next Generation as our source content and our 1x1 pixel image as the baseline.

Image Size	JPEG Bytes	PNG Bytes	JPEG Overhead Percentage	PNG Overhead Percentage
1 x 1	632	69	100.0%	100.0%
64 x 64	2019	8532	31.30%	0.81%
128 x 128	4912	30724	12.87%	0.22%
256 x 256	14267	111642	4.43%	0.06%
512 x 512	43424	410782	1.46%	0.017%
1024 x 1024	135570	1515218	0.47%	0.0046%
2048 x 2048	423298	5528685	0.15%	0.0012%
4096 x 4096	1309545	19513354	0.048%	0.00035%
8192 x 8192	4549578	62798290	0.014%	0.00011%

Table 4.4 Comparison of JPEG vs PNG compression performance.

Clearly, we can reduce overhead by using very large images. But very large images introduce a new problem. It is unlikely that our users will be very satisfied waiting for a 8192 by 8192 image to download and display, especially when their monitors can show only 1024 by 768. They are able to view only 1.17% of the pixels in the image at one time. Also, very large images consume a lot of system memory and may not be usable at all on smaller or older devices.

There is another consideration to be made that is specific to JPEG images. The JPEG compression algorithm is block based. It commonly uses 16 by 16 blocks of pixels as minimum compression units. If an image's pixels are not evenly divisible by 16 in each dimension, it will pad the image with empty values. We can prove this by creating a series of JPEG images sized 1 to 500 pixels. Each image consists of all black pixels.

The distinct stair-step pattern in Figure 4.23 shows that the images increase in compressed size by 16 pixel increments. Therefore, we should choose tile sizes that are powers of 16, like 16, 32, 64, etc. This partially explains why our overhead calculations for JPEG and PNG images showed that overhead as a percentage from

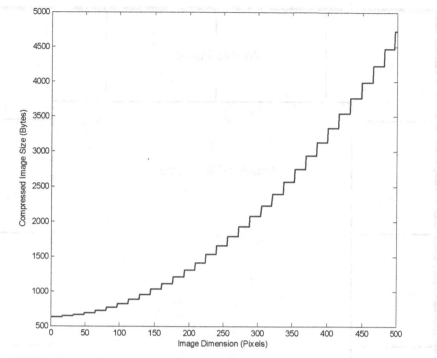

Fig. 4.23 Graph shows the step-pattern for size of a JPEG-compressed image.

JPEG images is much larger than PNG images. The 1 by 1 JPEG image would be the same size as a 16 by 16 JPEG image. This would not be the case for PNG images.

To determine the actual appropriate tile size, we can create an optimization function. We want to minimize both the number of individual images required to virtualize the map view and the number of wasted pixels. Wasted pixels are pixels that are transmitted and decoded but not part of the virtualized map view (See Figure 4.24).

The best way to minimize wasted pixels would be to make all of our tiles 1 by 1, and then we would never have to decode any pixels that are not part of the final image. However, the overhead of having to retrieve and decode thousands and thousands of image files per map view would make our system unusable.

We can experimentally determine the proper tile size for our system. First we need to guess the typical size of a virtualized map view. For this example, we will use 1024 by 768 pixels. Given this size, we can generate a large number of random map views for a given scale. For each of those random map views, we will calculate the number of tiles needed to fill that that view and the number of wasted pixels. We will perform this calculation for all the tile dimensions that we are considerating using. For this example, we will use tile dimensions 16, 32, 64, 128, 256, 512, 1024, and 2048.

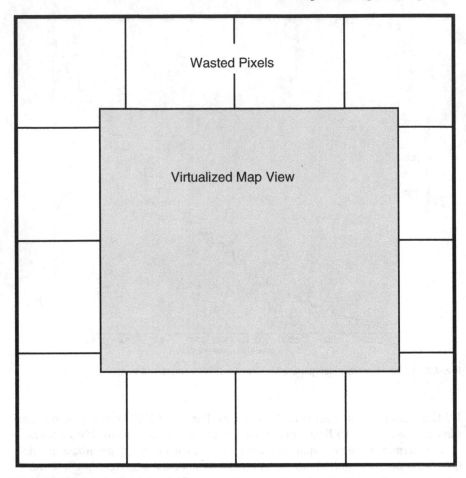

Fig. 4.24 Wasted pixels are decoded but not used as part of the virtualized map view.

For an example map scale, we will use scale 10 from the logical tile scheme that we developed in Chapter 2. Recall, that scale 10 is simply defined as having 2^{10} (1024) columns by 2^9 (512) rows. Each tile is $\frac{1024}{360.0}$ (2.8$\overline{4}$) degrees wide and long.

Generating random map views is fairly easy. Since all map views have the same aspect ratio, we need to generate a large number of random center locations. The center locations would be in the range of -180 to 180 for the longitude coordinate and -90 to 90 for the latitude coordinate. Then, for each tile size, we can extrapolate the map view bounds from the center location.

When generating our random map views, we have to consider that cases in which some portion of our map view will go beyond the normal bounds of the Earth. For example, the longitudes might be greater than 180.0 or less than -180.0. There are two ways to deal with this. First we can constrain our randomization function to a range of coordinates that is guaranteed to never generate map view bounds outside our given range, or, secondly, we could simply perform those calculations without caring if the boxes overlap our acceptable coordinates. We will choose the latter method. If a randomly computed map box strays beyond the -180 to 180 and -90 to 90 bounds, we will compute wasted pixels and tiles that were accessed as if there were tiles and pixels in those areas. This is a practical decision because many map clients perform wrapping in boundary areas. They pull images and pixels from the other side of the map to fill in the boundary overlaps.

The algorithm in Listing 4.12 generates 10,000 randomized map center locations for scale 10 and tile dimensions of 16, 32, 64, 128, 256, 512, 1024, and 2048. It computes the total number of tiles accessed and the wasted pixels for each access. Those results are shown in Table 4.5.

Tile Size	Number of Tiles Accessed	Wasted Pixels
16	3185.0	28928.0
32	825.0	58368.0
64	221.0	118784.0
128	63.0	245760.0
256	20.0	524288.0
512	7.5021	1180198.5024
1024	3.4938	2877082.8288
2048	2.0677	7886130.3808

Table 4.5 Tiles accessed and wasted pixels for 1024 by 768 map view. 10,000 random map views averaged.

When the results are plotted, it is easy to see the optimal point, as shown in Figure 4.25. We have normalized the tiles accessed and pixels wasted values. The two lines cross very near to when the tiles are sized 128 by 128. This statistic might lead us to select tiles sized 128 by 128. However, these calculations are performed in pixel counts. We are disregarding the important computations performed earlier to determine overhead percentages for each tile size. Re-computing the optimization and substituting pixels wasted with total bytes accessed yields a different result. Furthermore, the result can be plotted with just one line to see the bytes used as a function of tile size; see Figure 4.26. Table 4.6 shows our results using the listed tile image sizes in bytes.

Tile Size	Standardized Image Size in Bytes	Number of Tiles Accessed	Total Bytes Accessed
16	759	3185.0	2417415.0
32	1062	825.0	876150.0
64	2019	221.0	446199.0
128	4912	63.0	309456.0
256	14267	20.0	285340.0
512	43424	7.5	326070.816
1024	135570	3.5	474305.202
2048	423298	2.06	873687.072

Table 4.6 Bytes accessed for different sized tiles.

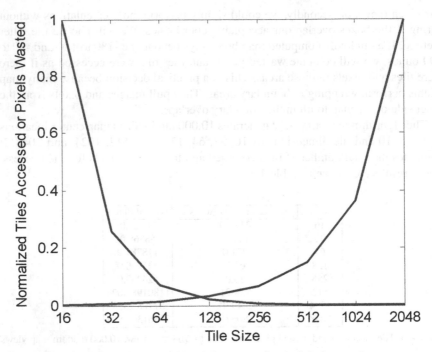

Fig. 4.25 Graph of the normalized number of tiles accessed and of pixels wasted.

Clearly the 16 by 16 tiles are very inefficient. They require the most bytes to be read, even though our earlier computations showed that they generated the fewest wasted pixels. The effect of the wasted pixels is seen as the tile sizes get larger. According to this graphic, tiles sizes 128, 256 or 512 are all close to optimal.

What if we consider more than one map view resolutions? Up to now, we have considered only 1024 by 768 map view resolutions. Figure 4.27 shows the results for map resolutions 640 by 480, 800 by 600, 1024 by 768, 1280 by 960, 1400 by 1050, and 1600 by 1200.

The results are similar: we still see the bottom (or optimum area) of our plots around the 128, 256, and 512 area. Figure 4.28 shows the results for PNG image's sizes instead of JPEG sizes. We can see the effect of reduced overhead in PNG images, but otherwise the plots are similar.

Figure 4.29 shows the JPEG bytes accessed plotted as differences from one tile size to the other. In this figure we can see that the line is almost flat from 256 to 512. This indicates that there is very little difference between these two tile sizes in terms of total bytes accessed.

Since we have moved from considering pixels to compressed image bytes, we should also consider computation time required to decompress the compressed tile images. Table 4.7 shows the average decode times for tiles of varied sizes in both

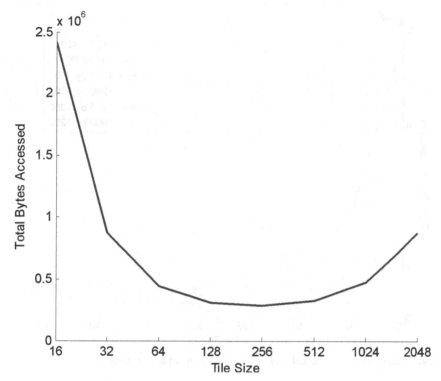

Fig. 4.26 Graph of bytes accessed vs. tile size.

JPEG and PNG formats. If we put these numbers into our previous optimization plots we get Figure 4.30 and Figure 4.31. In both of these experiments, we see that our plot has a minimum around the 512 tile size.

Tile Size	Decode Time JPEG (Milliseconds)	Decode Time PNG (Milliseconds)
16	2.5	2.34
32	2.5	2.5
64	2.66	3.12
128	3.91	4.22
256	5.0	7.97
512	10.94	21.1
1024	31.56	69.22
2048	113.75	258.44

Table 4.7 Decode times for JPEG and PNG tiles.

Further enhancements to this approach are possible. We have considered only random map views that exactly match our pre-determined map scales. In practice this will occur only for map clients that adhere to those fixed map scales. In addition, we have fixed the map scale by number of tiles and varied the tile size. In practical

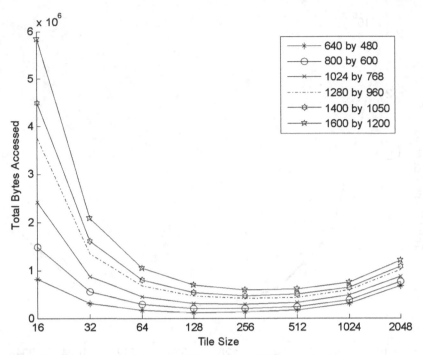

Fig. 4.27 Graph of bytes accessed using JPEG tiles for multiple map display resolutions.

terms this means that while each tile covers that same portion of the earth, the real map resolution of each tile varies with its size. Both of these shortcomings can be addressed by replacing our fixed map scale of 10 with a randomly selected map scale. The randomly selected map scale should be chosen from a continuous range instead of fixed discrete scales. In these cases, we will have to scale the pixels from the covered tiled region to match the scale of the map view with the randomly chosen scale.

In conclusion, from consideration of the results given above we are going to use 512×512 for the tile sizes in this book. Our analyses indicate the 256×256 would also be a good choice. However, we should consider a final point. It takes four 256×256 tiles to cover the area of one 512×512 tile. Thus when we create a large number of tiles, if we use 256×256 tiles, we will have four times the database entries or four times the tile image files, and either way our indexes will be four times the size. As we cover techniques for producing and storing tiles, we will see that these are significant costs.

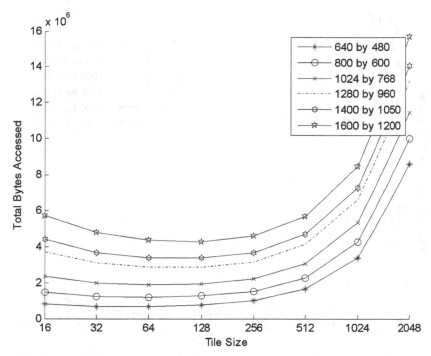

Fig. 4.28 Graph of bytes accessed using PNG tiles for multiple map display resolutions

4.6 Tuning Image Compression

In dealing with very large tiled image sets, it may be important to try to reduce the amount of storage space required. The compression quality of JPEG compressed images can be adjusted to produce smaller or larger compressed files. As the compression rate is increased the file size decreases. Many software platforms support setting the JPEG quality ratio to values in a pre-defined range; these can be either 0 to 1, or 0 to 100. Higher quality values mean less compression. If we apply varied quality settings to a 512 by 512 JPEG image taken from the Blue Marble data set, we get the file size differences shown in Table 4.8.

Quality Setting	JPEG File Size in Bytes
90	62128
80	45475
70	42141
60	38746
40	21967
10	9952

Table 4.8 File sizes for different JPEG quality settings.

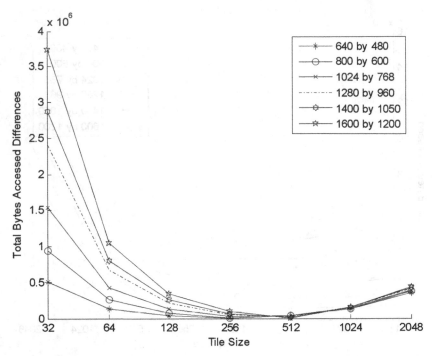

Fig. 4.29 Change in the number of bytes accessed as tile size increases (JPEG tiles).

Figure 4.32 is a tiled image from the Blue Marble set. Figures 4.33(a), 4.33(b), 4.33(c), 4.33(d), 4.33(e), and 4.33(f) show the results of applying the various compression quality settings.

The 80 and 90 quality settings are visually hard to distinguish, but lower values show compression artifacts. For example, the 10 quality image is quite blurry. There is a significant drop in storage space required from quality setting 90 to 80. After that, the drops are less pronounced. Anyone producing large tile sets should take time to manually set the quality setting appropriate to their application. Tile producers might consider using a lower quality setting for lower resolution tiles and using a higher setting for higher resolution tiles. This would provide tile system users with lower quality overview images, but the option to zoom in for higher quality map views.

Even though the PNG format is technically lossless, we can apply some lossy techniques to reduce the file sizes of PNG images. Recall that the PNG format is sensitive to the number of colors used in an image. If we can reduce the number of colors in an image, a process called "color quantization," we can reduce the size of the compressed PNG file.

There are many algorithms for color quantization. A very simple algorithm would simply reduce the byte space available for colors. So instead of an RGB image with 8 bits for each color channel, we could only allow 7 bits for each channel. Listing 4.8 shows this algorithm applied to an example map image.

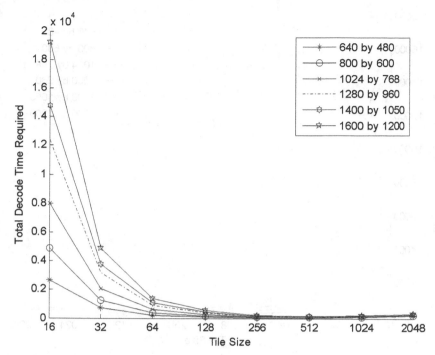

Fig. 4.30 Decode time for JPEG images as tile size increases.

Listing 4.8 Simple algorithm for reducing the color space of an image.

```
byte[] squeezeColors(BufferedImage bi, int bitsPerColor) throws IOException {
        int bitsToShift = 8 - bitsPerColor;
        for (int i = 0; i < bi.getWidth(); i++) {
            for (int j = 0; j < bi.getHeight(); j++) {
                Color c = new Color(bi.getRGB(i, j));
                int b = c.getBlue();
                int g = c.getGreen();
                int r = c.getRed();
                b = (b >> bitsToShift) << bitsToShift;
                g = (g >> bitsToShift) << bitsToShift;
                r = (r >> bitsToShift) << bitsToShift;
                Color c2 = new Color(r, g, b);
                bi.setRGB(i, j, c2.getRGB());
            }
        }
        ByteArrayOutputStream baos = new ByteArrayOutputStream();
        ImageIO.write(bi, "png", baos);
        byte[] data = baos.toByteArray();
        return data;
    }
```

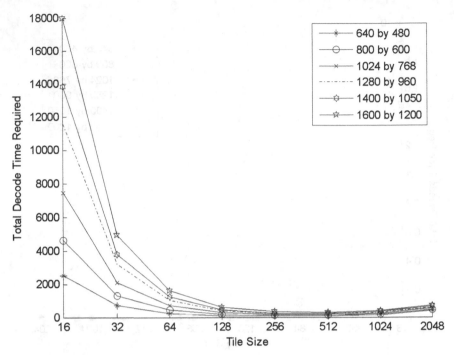

Fig. 4.31 Decode time for PNG images as tile size increases.

```
b = (b >> bitsToShift) << bitsToShift;
g = (g >> bitsToShift) << bitsToShift;
r = (r >> bitsToShift) << bitsToShift;
```

The key part of the code is the following section:

We shift each color component to the right and then back to the left. This has the effect of rounding off or zeroing out the right most or least significant bits for each component.

Table 4.9 shows the effect of this simple algorithm on the number of colors and the resulting file sizes on the image shown in Figure 4.16.

Reducing the color palette from 24 bits to 21 bits reduces the number of colors from 2,372 to 2,094, and it delivers a significant reduction in file size, nearly 45%. Figures 4.34(a), 4.34(b), 4.34(c), 4.34(d), 4.35(a), 4.35(b), and 4.35(c) show the effect of reducing the color palette. Only the most severe reductions provide a visible decrease in the image's quality.

There are more sophisticated algorithms. The "Quantize" algorithm in the ImageMagick software package [3] uses a tree structure to classify and reduce the number of colors in an image. Rather than simply reducing the bit depth as we did in

---

[3] http://www.imagemagick.org/script/quantize.php

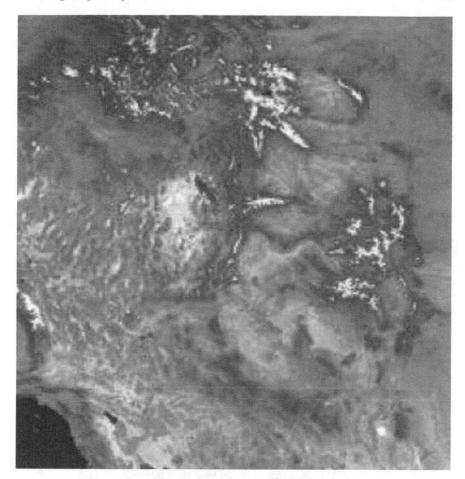

Fig. 4.32 Blue Marble tile image.

RGB Bits	Image Colors	File Size (Bytes)
24	2,372 (Original)	329,662
21	2,094	182,413
18	1,728	176,675
15	1,217	163,290
12	624	148,144
9	191	102,665
6	43	48,660
3	8	26,249

Table 4.9 Results of Simple Color Reduction Algorithm.

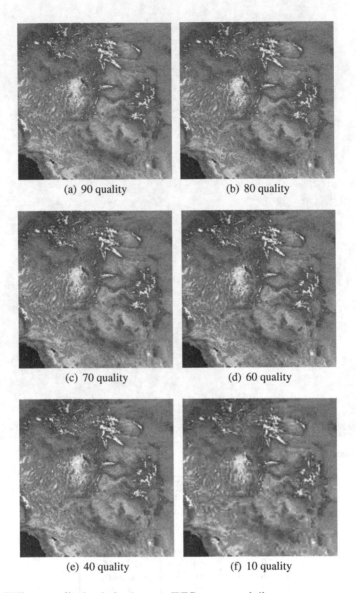

(a) 90 quality   (b) 80 quality

(c) 70 quality   (d) 60 quality

(e) 40 quality   (f) 10 quality

Fig. 4.33 Different quality levels for the same JPEG compressed tile.

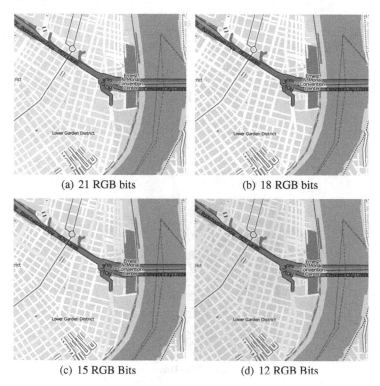

(a) 21 RGB bits

(b) 18 RGB bits

(c) 15 RGB Bits

(d) 12 RGB Bits

Fig. 4.34 Comparison between PNG compressed images with differening numbers of bits to represent color. (Part 1)

tour example, this algorithm attempts to intelligently discard colors. It functions by minimizing the color reduction error by calculating the error resulting from the elimination of any single color and then eliminating the colors that have the least impact on the overall error.

Image Colors	File Size (Bytes)
2,372 (Original)	329,662
1659	349,132
798	298,533
400	286,660
100	209,952
50	202,308
16	165,902
8	104,230

Table 4.10 Color reductions using the Quantize algorithm

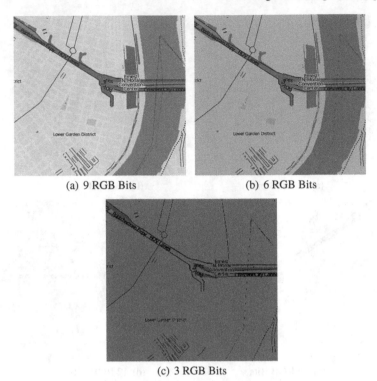

(a) 9 RGB Bits   (b) 6 RGB Bits

(c) 3 RGB Bits

Fig. 4.35 Comparison between PNG compressed images with different numbers of bits to represent color. (Part 2)

Table 4.10 shows the color reduction to file size reduction numbers. Figures 4.36(a), 4.36(b), 4.36(c), and 4.36(d) show the images reduced to 100, 50, 16, and 8 colors respectively. The images are still usable with as few as 100 colors, but the compression improvements are less significant. Reducing from 2,372 to 100 colors yields only a 36% reduction. We also note the peculiar result that reducing from 2,372 colors to 1,659 actually yields an increase in the file size.

Those producing tiled images will need to conduct their own trials to determine if color reduction is a useful step in their tile production process. We should note that we are reducing the colors of rendered map graphics after they have been rendered. It would be more efficient and more sensible to simply render them with fewer colors from the beginning, if possible.

There are other techniques to reduce the sizes of compressed images. For example, we could apply smoothing filters to the images. However, this type of technique is only effective because it reduces the amount of information stored in the image. This also reduces the usefulness of the image.

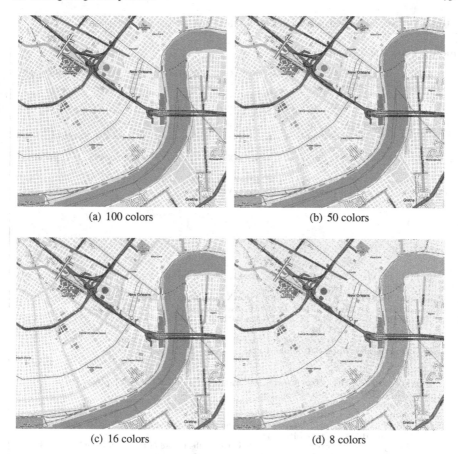

(a) 100 colors   (b) 50 colors

(c) 16 colors   (d) 8 colors

Fig. 4.36 Comparison between PNG compressed images with differening numbers of total colors in the image.

Listing 4.9 Java implementations of nearest neighbor and bilinear interpolation.

```java
public class ImageScaling {

    public static void scale(BufferedImage s, BufferedImage t, BoundingBox sbb,
            BoundingBox tbb) {
        int target_height = t.getHeight();
        int target_width = t.getWidth();
        for (int j = 0; j < target_height; j++) {
            for (int i = 0; i < target_width; i++) {
                Point2DDouble point = geolocate(tbb, i, j, target_height,
                    target_width);
                if (!((point.x > sbb.maxX) || (point.y > sbb.maxY) || (point.x
                    < sbb.minX) || (point.y < sbb.minY))) {
                    int pixelval = interpolate(s, sbb, point);
                    t.setRGB(i, j, pixelval);
                }
            }
        }
    }



    public static Point2DDouble geolocate(BoundingBox coords, int i, int j, int
            width, int height) {
        double pixel_width = (coords.maxX - coords.minX) / width;
        double pixel_height = (coords.maxY - coords.minY) / height;
        double x = (i + 0.5) * pixel_width + coords.minX;
        int adj_j = height - j - 1;
        double y = (adj_j + 0.5) * pixel_height + coords.minY;
        return new Point2DDouble(x, y);
    }

    private static int interpolate_nn(BufferedImage s, BoundingBox sbb,
            Point2DDouble point) {
        double tx = point.x;
        double ty = point.y;
        int i = (int) Math.round((tx - sbb.minX) / (sbb.maxX - sbb.minX) * (
                double) s.getWidth());
        int j = s.getHeight() - 1 - ((int) Math.round((ty - sbb.minY) / (sbb.
                maxY - sbb.minY) * (double) s.getHeight()));
        return s.getRGB(i, j);
    }

    private static int interpolate_bl(BufferedImage s, BoundingBox sbb,
            Point2DDouble point) {
        double tx = point.x;
        double ty = point.y;
        int source_height = s.getHeight();
        int source_width = s.getWidth();
        double temp_x = (tx - sbb.minX) / (sbb.maxX - sbb.minX) * source_width;
        double temp_y = source_height - 1 - ((ty - sbb.minY) / (sbb.maxY - sbb.
                minY) * source_height);
        int i = (int) Math.floor(temp_x);
        int j = (int) Math.floor(temp_y);
        double weight_x = temp_x - i;
        double weight_y = temp_y - j;
        if (j == source_height) {
            j = j - 1;
        }
        if (i == source_width) {
            i = i - 1;
        }
        if (j < 0) {
            j = 0;
        }
        if (i < 0) {
```

```
58        i = 0;
59    }
60    int val_0_0 = s.getRGB(i, j);
61    int val_0_1 = s.getRGB(i, j);
62    int val_1_0 = s.getRGB(i, j);
63    int val_1_1 = s.getRGB(i, j);
64    int pixel_val_r = getPixelValue(rmask(val_0_0), rmask(val_0_1), rmask(
          val_1_0), rmask(val_1_1), weight_x, weight_y);
65    int pixel_val_g = getPixelValue(gmask(val_0_0), gmask(val_0_1), gmask(
          val_1_0), gmask(val_1_1), weight_x, weight_y);
66    int pixel_val_b = getPixelValue(bmask(val_0_0), bmask(val_0_1), bmask(
          val_1_0), bmask(val_1_1), weight_x, weight_y);
67    int pixel_val = pixel_val_r << 16 | pixel_val_g << 8 | pixel_val_b | 0
          xff000000;
68    return pixel_val;
69    }
70
71    private static int bmask(int val) {
72        int b = val & 0x000000ff;
73        return b;
74    }
75
76    private static int gmask(int val) {
77        int b = (val >> 8) & 0x000000ff;
78        return b;
79    }
80
81    private static int rmask(int val) {
82        int r = (val >> 16) & 0x000000ff;
83        return r;
84    }
85
86    private static int getPixelValue(int val_0_0, int val_0_1, int val_1_0, int
          val_1_1, double weight_x, double weight_y) {
87        int pixel_val = (int) ((1 - weight_x) * (1 - weight_y) * val_0_0 +
              weight_x * (1 - weight_y) * val_0_1 + (1 - weight_x) * (weight_y)
88                               * val_1_0 + weight_x * weight_y * val_1_1);
89        return pixel_val;
90    }
91
```

Many programming environments provide built-in tools for scaling and subsetting images. This changes our algorithms slightly. Instead of performing pixel by pixel calculations, we compute a single set of transformation parameters and pass those to the built in image manipulation routines. The following code sections show how to use those built in routines in Java and Python.

Java Image Scaling and Subsetting

```
97    public static void drawImageToImage(BufferedImage source, BoundingBox
          source_bb, BufferedImage target, BoundingBox target_bb) {
98        double xd = target_bb.maxX - target_bb.minX;
99        double yd = target_bb.maxY - target_bb.minY;
100       double wd = (double) target.getWidth();
101       double hd = (double) target.getHeight();
102       double targdpx = xd / wd;
103       double targdpy = yd / hd;
104       double srcdpx = (source_bb.maxX - source_bb.minX) / source.getWidth();
105       double srcdpy = (source_bb.maxY - source_bb.minY) / source.getHeight();
106       int tx = (int) Math.round(((source_bb.minX - target_bb.minX) / targdpx)
              );
107       int ty = target.getHeight() - (int) Math.round(((source_bb.maxY -
              target_bb.minY) / yd) * hd) - 1;
108       int tw = (int) Math.ceil(((srcdpx / targdpx) * source.getWidth()));
109       int th = (int) Math.ceil(((srcdpy / targdpy) * source.getHeight()));
110       Graphics2D target_graphics = (Graphics2D) target.getGraphics();
```

```
111
112        //use one of these three statements to set the interpolation method to
             be used
113        target_graphics.setRenderingHint(RenderingHints.KEY_INTERPOLATION,
             RenderingHints.VALUE_INTERPOLATION_NEAREST_NEIGHBOR);
114        target_graphics.setRenderingHint(RenderingHints.KEY_INTERPOLATION,
             RenderingHints.VALUE_INTERPOLATION_BILINEAR);
115        target_graphics.setRenderingHint(RenderingHints.KEY_INTERPOLATION,
             RenderingHints.VALUE_INTERPOLATION_BICUBIC);
116
117        target_graphics.drawImage(source, tx, ty, tw, th, null);
118    }
```

Listing 4.10 Java image scaling and subsetting.

```
1  public static void drawImageToImage(BufferedImage source,
2            BoundingBox source_bb, BufferedImage target,
3            BoundingBox target_bb) {
4      double xd = target_bb.maxX - target_bb.minX;
5      double yd = target_bb.maxY - target_bb.minY;
6      double wd = (double) target.getWidth();
7      double hd = (double) target.getHeight();
8      double targdpx = xd / wd;
9      double targdpy = yd / hd;
10     double srcdpx = (source_bb.maxX - source_bb.minX) / source.getWidth();
11     double srcdpy = (source_bb.maxY - source_bb.minY) / source.getHeight();
12     int tx = (int) Math.round(((source_bb.minX - target_bb.minX) / targdpx));
13     int ty = target.getHeight() - (int) Math.round(((source_bb.maxY - target_bb
            .minY) / yd) * hd) - 1;
14     int tw = (int) Math.ceil(((srcdpx / targdpx) * source.getWidth()));
15     int th = (int) Math.ceil(((srcdpy / targdpy) * source.getHeight()));
16     Graphics2D target_graphics = (Graphics2D) target.getGraphics();
17
18     //use one of these three statements to set the interpolation method to be
            used
19     target_graphics.setRenderingHint(RenderingHints.KEY_INTERPOLATION,
            RenderingHints.VALUE_INTERPOLATION_NEAREST_NEIGHBOR);
20     target_graphics.setRenderingHint(RenderingHints.KEY_INTERPOLATION,
            RenderingHints.VALUE_INTERPOLATION_BILINEAR);
21     target_graphics.setRenderingHint(RenderingHints.KEY_INTERPOLATION,
            RenderingHints.VALUE_INTERPOLATION_BICUBIC);
22
23     target_graphics.drawImage(source, tx, ty, tw, th, null);
24  }
```

Listing 4.11 Python image scaling and subsetting.

```
1  import Image, ImageDraw # requires the Python Imaging Library (PIL) addon to
        python
2
3  def drawImageToImage(source, sourceBoundingBox, target, targetBoundingBox):
4      # calculations to determine the degrees per pixel for each dimension of the
            target image
5      targetXDelta = targetBoundingBox.maxX - targetBoundingBox.minX
6      targetYDelta = targetBoundingBox.maxY - targetBoundingBox.minY
7      targetWidth = target.size[0]
8      targetHeight = target.size[1]
9      targetDegPerPixelX = targetXDelta / float(targetWidth)
10     targetDegPerPixelY = targetYDelta / float(targetHeight)
11
12     # calculations to determine the degrees per pixel for each dimension of the
            source image
13     # (we collapse the equations into two lines)
14     sourceXDelta = sourceBoundingBox.maxX - sourceBoundingBox.minX
15     sourceYDelta = sourceBoundingBox.maxY - sourceBoundingBox.minY
```

```
16    sourceWidth = source.size[0]
17    sourceHeight = source.size[1]
18    sourceDegPerPixelX = sourceXDelta / float(sourceWidth)
19    sourceDegPerPixelY = sourceYDelta / float(sourceHeight)
20
21    targetX = int(round((sourceBoundingBox.minX - targetBoundingBox.minX) /
          targetDegressPerPixelX))
22    targetY = targetHeight - int(round(((sourceBoundingBox.maxY -
          targetBoundingBox.minY) / targetYDelta) * targetHeight)) - 1
23    tw = int(math.ceil(((sourceDegPerPixelX / targetDegPerPixelX) * sourceWidth
          )))
24    th = int(math.ceil(((sourceDegPerPixelY / targetDegPerPixelY) *
          sourceHeight)))
25
26    # use one of these to set the interpolation method when resizing the target
          image
27    interpolation = Image.NEAREST
28    interpolation = Image.BILINEAR
29    interpolation = Image.BICUBIC
30
31    resizedSource = s.resize((tw,th), interpolation)
32
33    im.paste(resizedSource, (targetX, targetY, tw, th))
```

Listing 4.12 Generating randomized map view locations.

```
1  public static void tileSizeOptimization1() {
2      int numlocations = 10000;
3      int[] tilesizes = new int[] {16, 32, 64, 128, 256, 512, 1024, 2048};
4
5      int scale = 10;
6      int viewWidth = 1024;
7      int viewHeight = 768;
8
9      int numpoints = numlocations;
10     Point2DDouble[] randomPoints = getRandomPoints(numpoints);
11     int[] totalTilesAccessed = new int[tilesizes.length];
12     long[] totalPixelsWasted = new long[tilesizes.length];
13     for (int i = 0; i < tilesizes.length; i++) {
14         BoundingBox[] bb = getRandomMapViews(randomPoints, scale, tilesizes
               [i], viewWidth, viewHeight);
15         int[] tilesAccessed = getTilesAccessed(bb, scale);
16         long[] pixelsWasted = getWastedPixels(tilesAccessed, scale,
               viewWidth, viewHeight, tilesizes[i]);
17         totalTilesAccessed[i] = 0;
18         for (int j = 0; j < numlocations; j++) {
19             totalTilesAccessed[i] += tilesAccessed[j];
20             totalPixelsWasted[i] += pixelsWasted[j];
21         }
22     }
23
24     for (int i = 0; i < totalTilesAccessed.length; i++) {
25         System.out.println(totalTilesAccessed[i] / 10000.0);
26     }
27     for (int i = 0; i < totalTilesAccessed.length; i++) {
28         System.out.println(totalPixelsWasted[i] / 10000.0);
29     }
30
31  }
32
33 //this method determines the number of pixels that are decoded from the number
       of tiles //accessed and subtracts the number of pixels in the current map
       view
34 private static long[] getWastedPixels(int[] tiles, int scale, int viewWidth,
       int viewHeight, int tilesize) {
35     long[] pixels = new long[tiles.length];
```

```
36            int pixelsPerTile = tilesize * tilesize;
37            int pixelsPerView = viewWidth * viewHeight;
38            for (int j = 0; j < pixels.length; j++) {
39                pixels[j] = (tiles[j] * pixelsPerTile) - pixelsPerView;
40                if (pixels[j] < 0) {
41                    System.out.println(pixelsPerView + ":" + tiles[j] + ":" +
                         pixelsPerTile);
42                }
43            }
44            return pixels;
45        }
46
47        //this calculates the number of tiles that are needed to cover each
              bounding box
48        public static int[] getTilesAccessed(BoundingBox[] boxes, int scale) {
49            int[] tiles = new int[boxes.length];
50            for (int i = 0; i < tiles.length; i++) {
51                long mincol = (long) Math.floor(getColForCoord(boxes[i].minX, scale
                     ));
52                long minrow = (long) Math.floor(getRowForCoord(boxes[i].minY, scale
                     ));
53                long maxcol = (long) Math.floor(getColForCoord(boxes[i].maxX, scale
                     ));
54                long maxrow = (long) Math.floor(getRowForCoord(boxes[i].maxY, scale
                     ));
55                tiles[i] = (int) ((maxcol - mincol + 1) * (maxrow - minrow + 1));
56            }
57            return tiles;
58
59        }
60        //this returns the tile column coordinate that contains the longitude "x"
              for scale "scale"
61        static double getColForCoord(double x, int scale) {
62            double coord = x + 180.0;
63            coord = coord / (360.0 / Math.pow(2.0, (double) scale));
64            return (coord);
65        }
66        //this returns the tile column coordinate that contains the latitude "y"
              for scale "scale"
67        static double getRowForCoord(double y, int scale) {
68            double coord = y + 90.0;
69            coord = coord / (360.0 / Math.pow(2.0, (double) scale));
70            return (coord);
71        }
72        //this computes "numpoints" random x and y locations within our map
              coordinate system
73        private static Point2DDouble[] getRandomPoints(int numpoints) {
74            Point2DDouble[] points = new Point2DDouble[numpoints];
75            for (int i = 0; i < numpoints; i++) {
76                double centerX = Math.random() * 360.0 - 180.0;
77                double centerY = Math.random() * 180.0 - 90.0;
78                Point2DDouble point = new Point2DDouble(centerX, centerY);
79                points[i] = point;
80            }
81            return points;
82        }
83        //the extrapolates our random x and y locations into map view boxes
84        public static BoundingBox[] getRandomMapViews(Point2DDouble[] centerPoints,
                 int scale, int tilesize, int viewWidth, int viewHeight) {
85            //these two are always the same
86            double tileWidthDegrees = 360.0 / Math.pow(2, scale);
87            double tileHeightDegrees = 180.0 / Math.pow(2, scale - 1);
88
89            double degreesPerPixel = tileWidthDegrees / tilesize;
90            double viewWidthDegrees = viewWidth * degreesPerPixel;
91            double viewHeightDegrees = viewHeight * degreesPerPixel;
92
```

```
 93    BoundingBox[] boxes = new BoundingBox[centerPoints.length];
 94    for (int i = 0; i < boxes.length; i++) {
 95        double centerX = centerPoints[i].x;
 96        double centerY = centerPoints[i].y;
 97        double minx = centerX - viewWidthDegrees / 2.0;
 98        double maxx = centerX + viewWidthDegrees / 2.0;
 99        double miny = centerY - viewHeightDegrees / 2.0;
100        double maxy = centerY + viewHeightDegrees / 2.0;
101        BoundingBox bb = new BoundingBox(minx, miny, maxx, maxy);
102        boxes[i] = bb;
103    }
104    return boxes;
105  }
```

References

1. Denning, P., Schwartz, S.: Properties of the working-set model. Communications of the ACM **15**(3), 198 (1972)
2. Keys, R.: Cubic convolution interpolation for digital image processing. IEEE Transactions on Acoustics, Speech and Signal Processing **29**(6), 1153–1160 (1981)
3. Press, W., Teukolsky, S., Vetterling, W., Flannery, B.: Numerical recipes in C. Cambridge university press Cambridge (1992)

Chapter 5
Image Tile Creation

The previous chapter explained the techniques for manipulating geospatial images. This chapter will build on those techniques to explain how a system can be constructed to create sets of tiled geospatial images. In general terms, there are two types of tile generation systems: those that pre-render tiled images and those that render the images in direct response to user queries. Pre-rendering the tiles can require significant processing time, including processing tiles that may never be viewed by users. Rendering tiles just-in-time can save setup time but may require users to wait longer for requested maps. Beyond these differences there are significant technical reasons that usually force us to choose one type of system over the other.

Systems that serve tiled geospatial images from rendered vector content almost always use a form of just-in-time tiling. There are three reasons for this:

- Storage space: Rendered image tiles require a significant amount of storage space relative to vector map content. A collection of geospatial features might be 100 megabytes in vector form but could grow to several terabytes when rendered over several different levels.
- Processing time: Pre-rendering image tiles requires a significant amount of time, and many of those tiles may be in geographic areas of little interest to users. The most efficient method of deciding what tiles to render is to wait until they are requested by actual users.
- Overview images: Overview images, i.e., very low zoom level images, can be rendered directly from geospatial vectors. Unlike raster based tile systems, there is no need to render the high level views first and then generate scaled down versions.

Conversely, tiling systems that primarily draw from sets of geospatial imagery typically pre-render all image tiles. There are two reasons for this:

- Processing time: Reformatting, scaling, and reprojecting of imagery are often required in the tile creation process. These steps can be too time consuming for users to wait for in real time.

J.T. Sample and E. Ioup, *Tile-Based Geospatial Information Systems:*
Principles and Practices, DOI 10.1007/978-1-4419-7631-4_5,

- Overview images: Low level images must be created from higher resolution versions of the same imagery. This requires the higher levels to be completely rendered before the low resolution levels can be completed.

Since the primary focus of this book is tiling systems based on imagery, this chapter will examine tiling systems that pre-render image tiles. Chapter 11 will discuss tiling systems based on vector geospatial data.

5.1 Tile Creation from Random Images

Tiled image sets are created from collections of random source images. We call the source images random because, unlike our tiled images, the source images may have sizes and boundaries that follow no specific system. Collections of source images come in varied forms. For example, we might have high-resolution aerial imagery of the world's 50 largest cities. Each city is represented by a small number (5 to 50) of large source images (approximately 10,000 by 10,000 pixels). Each of the source images can have a different size, cover a different portion of the earth, or have a different map resolution. Taken together, all of the source images for all of the cities form a single map layer. Common file formats for source images include GEOTIFF, MrSID, and GEO-JPEG2000. However, almost any image format can be used, as long as the geospatial properties of the image are encoded either in the image file or along side it.

At this point it is useful to define the concept of a map layer. Layers are typically the atomic unit for requesting map data from web-based geospatial systems. A map layer is a logical grouping of geospatial information. The term "layer" is used to convey the idea of multiple graphical layers stacked in some order in a visual display. Layers are formed by logical groupings of geospatial data. For example, an "entertainment" map layer would include locations of movie theaters, parks, zoos, museums, etc. In the case of imagery, we will formally define a layer as a single map view for a given geographic area.

Corallary 1 *For a single layer, there exists one and only one tile image for a specific tile address.*

For example, consider that we have aerial imagery over Alaska. The snow cover over Alaska varies from month to month. We might have source image sets for every month of the year. Each of those image sets covers the same area. For a given tile address, each image set would have a tiled image with different content. Therefore, we have to group them into separate layers, one layer for the image set for each month.

5.2 Tile Creation Preliminaries

For the purposes of this chapter, we will consider the problem of taking a set of random source images and converting them into a layer of tiled images. Since a set of random source images will have images of varying size and resolution, the tile creation process is simply the process of scaling and shaping image data from the source images into tile-sized pieces.

5.2.1 Bottom-Up Tile Creation

Each layer of tile images has multiple levels. A tile set starts with a base level; the base level is the level with the highest number and the highest resolution imagery. Each subsequent level is a lower version of the level preceding it. Figure 5.1 shows three levels of the same image layer. In this example, Level 3 is the base level. Levels 2 and 1 are lower resolution versions of the same data. The images in Figure 5.1 show the tile boundaries according to the logical tile scheme presented in Chapter 2.

Definition 1 *The base level for a tile layer is the highest resolution level, the level with the highest number for that tile layer.*

In the tile creation process, the base level is almost always completed, at least partially, before lower resolution levels. Therefore, we can say that tile creation is a bottom-up process in terms of map scale.

5.2.2 Choosing the Base Level for a Set of Source Images

Before a tile set can be created from a set of random images, the base level must be chosen. In some cases, a target base level is determined ahead of time. It could be required to integrate with other tile layers or client software. However, in most cases the base level is chosen to closely match the resolution of the sources images. A given set of random source images is unlikely to exactly match one of our predetermined level resolutions. So we must select the tile level that most closely matches our source images. Table 5.1 shows the first 19 levels in our logical tile scheme with 512 by 512 pixel tiles. It gives the number of horizontal and vertical tiles for each level along with each level's degrees per pixel (DPP). These are calculated using the formulas provided in Chapter 2. The DPP values will be used to choose base levels for sets of source images.

To perform this analysis, we will require some information from each of the source images in our set: image width and image height in pixels and minimum and maximum vertical and horizontal coordinates in degrees. As shown in Listing 5.1, we can compute the DPP value for our set of random images. We compute the DPP

Level 1

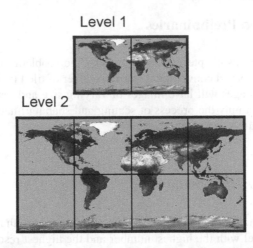

Level 2

Level 3

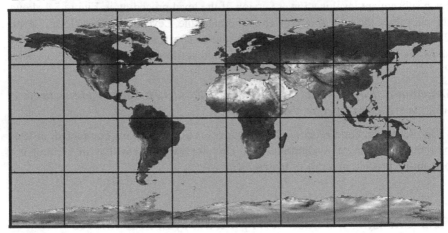

Fig. 5.1 Multiple zoom levels of the same layer.

value by combining the vertical and horizontal dimensions of the images. This is a valid procedure for tiled image projections that preserve the same DPP in each dimension as our logical tile scheme does. The calculations are performed in degrees. Source images stored in other projections might use meters with its coordinate system. In those cases, conversion to degrees is required. Suppose that for a set of source images, we have computed a DPP value of 0.03. This falls in between levels 4 and 5. If we choose level 4, we will be scaling DOWN our source images and thus losing a little bit of data from the source images. If we choose level 5, we will be scaling UP our source images. We will preserve all the data but take up more storage space.

For example, if our source image set takes up 10,000,000 bytes uncompressed with a DPP value of 0.03, when converted to level 4 it will take up 4,660,000 bytes,

Level	Horizontal Tiles	Vertical Tiles	Degrees Per Pixel
1	2	1	0.3515625
2	4	2	0.17578125
3	8	4	0.087890625
4	16	8	0.0439453125
5	32	16	0.02197265625
6	64	32	0.010986328125
7	128	64	0.0054931640625
8	256	128	0.00274658203125
9	512	256	0.001373291015625
10	1024	512	0.000686645507812
11	2048	1024	0.000343322753906
12	4096	2048	0.000171661376953
13	8192	4096	0.000085830688477
14	16384	8192	0.000042915344238
15	32768	16384	0.000021457672119
16	65536	32768	0.00001072883606
17	131072	65536	0.00000536441803
18	262144	131072	0.000002682209015
19	524288	262144	0.000001341104507

Table 5.1 Number of tiles and degrees per pixel for each level.

Listing 5.1 Computation of degrees per pixel for a set of random source images.

```
1  ddpX = 0.0
2  ddpY = 0.0
3  count = 0
4
5  for image in images:
6      count = count + 1
7      ddpX = ddpX + (image.maxX - image.minX)/image.width
8      ddpY = ddpY + (image.maxY - image.minY)/image.height
9
10 dppTotal = (dppX + dppY)/(2*count)
```

a reduction of 53%. When converted to level 5 it will take up 18,645,000 bytes, an increase of 86%, or nearly double the original amount.

Equation 5.1 gives an approximate computation of the storage space changes affected by transforming from the native level to a fixed tile level, where N is the native resolution in degrees per pixel, B is the base level resolution in degrees per pixel, S is the size of the source image, and R is the space required for the tiled image set. The exact storage space changes cannot be calculated analytically. There are several unknown factors, such as the impact of uneven source image breaks onto the tile boundaries. Since all images are stored in a compressed format, the exact storage space requirements can be calculated only by creating and compressing the images.

$$R = (\frac{N}{B}) * 2S \qquad (5.1)$$

In general, we want to choose the level with the closest DPP value that is lower than our native DPP as our base level. This will usually result in an increase of storage space requirements, but it will preserve the image information. Practically speaking, geospatial image data costs much more to create than to store. Satellites, aerial platforms, and cartographers are all more expensive than hard drives.

Since the tiled and source images are compressed, the actual increase in storage space requirements is usually smaller than Equation 5.1 would predict. Image compression algorithms attempt to compress images by storing just the information needed to reproduce the image. Since our rescaled tiled images are not adding any real image information, we can expect the compressed results to be similar in size to the original. Consider an example: the NASA Blue Marble image below is 2000 by 1000 pixels in size (Figure 5.2). Compressed as a JPEG image it is 201,929 bytes. If we resize the image to 3000 by 1500, we have increased the images number of pixels by a ratio of 2.25. However, the new larger image compressed as a JPEG takes up 360,833 bytes, a growth ratio of 1.79. So, the actual storage space requirements grew by 79%, not 125% as predicted by simple pixel calculations. Table 5.2 shows these results. It includes results for another scaled image that further illustrate the principle.

Once we have chosen and created the tiled images for the base level, the lower resolution levels can be created.

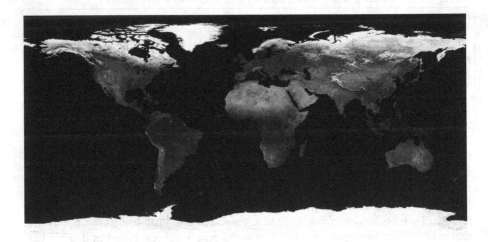

Fig. 5.2 2000 by 1000 Blue Marble image.

	Width	Height	Total Pixels	Percent Increased	Compressed Size	Percent Increased
Original Image	2000	1000	2,000,000		201,929	
Scaled Image 1	3000	1500	4,500,000	125%	360,833	79%
Scaled Image 2	5000	2500	12,500,000	525%	816,446	304%

Table 5.2 Compression ratios are greater for the same image at different resolutions.

Listing 5.2 Pull-based tile creation.

```
1 for t in tiles:
2     for s in sources:
3         if s.intersects(t):
4             p = extractPixels(s)
5             drawPixels(p,t)
```

Listing 5.3 Push-based tile creation.

```
1 for s in sources:
2     for t in tiles:
3         if t.intersect(s):
4             p = extractPixels(s)
5             drawPixels(p,t)
```

5.2.3 Pull-Based Versus Push-Based Tile Creation

There are two methods for creating the tiles from random source images: pull-based and push-based. Pull-based tile creation iterates over the desired tiles and pulls image data from the source images. Push-based tile creation iterates over the source images and pushes image data from them to the tiled images. There is little difference between these two approaches. The following pseudo code example shows that only the ordering of the iteration structure changes between the two methods, as shown in Listings 5.2 and 5.3.

In practice, there are several technical concerns that make the two methods substantially different. First and foremost is the issue of memory. If our computers had infinite memory, and all source and tile images could be held completely in memory, then there would be no effective difference between the two approaches. However, computers have limited memory, and we must move our source and tile images in and out of memory as we use them. Reading and decoding compressed source images from disk can be time consuming, as is compressing and writing tiled images to disk.

A second concern is that of multi-threading. Modern computers have multiple processing cores and can execute multiple threads simultaneously. A practical system must make use of multiple threads to be efficient, but it must also be careful to manage image resources in a thread safe fashion. Two threads should probably not operate on the same image tile at the same time.

For our first prototype tile creation system, we will use a pull-based method. Many of the tiles will contain data from multiple source images. If we iterate over source images first, as in a push-based method, then we will be swapping tiled images in and out of memory often. So those tiles will have to be swapped between memory and disk multiple times in the process of creating them. This poses several problems when it comes to tile storage, as many writes of small files or data blocks tends to cause fragmentation of a file system or database pages. In the next chapter we will discuss in more detail why it is important to write an image tile once and only once.

Unlike tiled images, our source images are used in a read-only fashion. We can safely swap them in and out of memory many times without having to perform any writes. This leads us to use a pull-based method. We will iterate over the tiles first and swap the large source images in and out of memory. This result may seem counterintuitive. Typically our source images are much larger than our tiled images. Sources images commonly range from 1,000 by 1,000 to 10,000 by 10,000. Our tile images are either 256 by 256 or 512 by 512. The large source images will take a significant amount of time to read and re-read from disk.

To mitigate this result, we will use a memory cache of source images. We will construct a Least Recently Used (LRU) cache of decoded source images in memory. LRU caches have a fixed size. If an element is added to an already full cache, the LRU cache will discard the least recently used element. Each time we access a source image, we will check if it is in the memory cache. If it is, then we do not have to read and decode the image. If the image is not in the memory cache, we will read and decode the image and place it in the cache.

The LRU cache works very well in this case. We will iterate over tiles in geographic order. Source images affect groups of tiles that border each other geographically. We can expect to have a high rate of "hits" on our memory image cache. The first tile that requests data from a source image will cause it to be loaded in the cache. The tiles immediately following the first tile will probably also use data from that source image which was just placed in the cache. This high cache hit rate provides a more efficient algorithm.

5.3 Tile Creation Algorithms

The following are the steps in the tile creation process:

1. Choose the base level for the tile set.
2. Determine the geographic bounds of the tile set. (This can be based on the bounds of the source images.)
3. Determine the bounds of the tile set in tile coordinates.
4. Initialize the tile storage mechanism.
5. Iterate over the tile set coordinates. For each tile, do the following:

   a. Compute the geographic bounds of the specific tile.

    b. Iterate over the source images. For each source image do the following:
       i. Determine if the specific source image intersects the tile being created.
      ii. If the source image and tile intersect,
         A. Check the cache for the source image. If it is not in the cache, load it
            from disk and save in the cache.
         B. Extract the required image data from the source image, and store it in
            the tiled image.
    c. Save the completed tiled image to the tile storage mechanism.

6. Clear the source image cache.
7. Finalize the tile storage mechanism.

Before presenting the computer code for executing these steps, we will define the following data types in Listing 5.5:

    BoundingBox: Wrapper for bounding rectangle in degrees.
    SourceImage: Wrapper for image dimensions and geographic bounds.
    TileAddress: Wrapper for a tile's row, column, and level coordinates.
    BufferedImage: Built-in Java class for memory images.
    TileOutputStream: Abstract class for output of tiled images.
    MemoryImageCache: Abstract class for a LRU cache of source images.

There are several key methods embedded in these data types. `TileAddress.getBoundingBox()` provides the bounding coordinates in degrees for an image tile address. `BoundingBox.intersects()` tests if two bounding boxes intersect each other. `BoundingBox.union()` is used to combine multiple bounding boxes into a single one. The abstract method `writeTile()` is used to provide a generic means for storing tiles. Concrete implementations of this will be discussed in the next chapter. Additional abstract methods, `getSourceImageData()` and `putSourceImageData()`, are used to provide access to the LRU source image cache. Implementation of this is left to the reader. We will also use the function `drawImageToImage`, which was defined in the previous chapter. Formulae for computing tile and geographic coordinates are derived in Chapter 2. We will use the constant `TILE_SIZE` to represent the width and height of our tiled images. This value is the same for the horizontal and vertical dimensions. See the previous chapter for a thorough discussion of how to choose the best tile size. Listing 5.6 is Java code for a basic, single threaded method for creating the base level of a tile set.

5.3.1 Scaling Process for Lower Resolution Levels

The previous algorithm created the base level. Next we create the lower resolution levels. Each lower level is based on the previous level. Because of the structured nature of our logical tile scheme, this process is much simpler than creation of the base level. Figure 5.3 shows that our lower resolution tiles are constructed directly and from exactly fours tiles from the previous level.

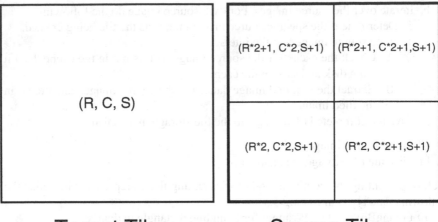

Fig. 5.3 Relationship of target tile for source tiles from previously computed level.

The basic algorithm is as follows:

1. For each level from (`base_level` - 1) to 1, do the following.
   a. Determine the bounds of the current tile level in tile coordinates.
   b. Initialize the tile storage mechanism.
   c. Iterate over the tile set coordinates. For each tile, do the following:
       i. Determine the four tiles from the higher level that contribute to the current tile.
       ii. Retrieve the four tile images or as many as exist.
       iii. Combine the four tile images into a single, scaled-down image.
       iv. Save the completed tiled image to the tile storage mechanism.
   d. Finalize the tile storage mechanism.

This algorithm uses the types defined in the previous section plus one additional type as defined in Listing 5.6. This type allows us to read the tiles from the previous levels. Additionally, an assumption is made that the TileInputStream and TileOutputStream in the algorithm are linked in some fashion. This allows us to write tiles in one stage and then read them out in the next stage. For example, when creating level 7, we will write level 7 tiles to the TileOutputStream. In the next stage, when we create level 6, we will have to read the level 7 tiles that we created in the previous step. Listing 5.7 shows the complete algorithm for creating the lower resolution layers from the base layer.

The process of creating tiled image sets from collections of random source images can be approached in a straightforward manner. In this chapter, we have detailed the basic process and algorithms for achieving this goal. We have built upon

Listing 5.4 Abstract class definition for the TiledInputStream.

```
1   abstract class TileInputStream {
2       abstract BufferedImage getTile(TileAddress address);
3   }
```

the image manipulation algorithms from previous sections. In the next chapter, we will present techniques for optimizing the creation of tiled image sets.

Listing 5.5 Java data types.

```
1  class BoundingBox {
2
3      double minx, miny, maxx, maxy;
4
5      public BoundingBox(double minx, double miny, double maxx, double maxy) {
6          this.maxx = maxx;
7          this.maxy = maxy;
8          this.minx = minx;
9          this.miny = miny;
10     }
11
12     boolean intersects(double minx, double miny, double maxx, double maxy) {
13         return !(minx > this.maxx || maxx < this.minx || miny > this.maxy ||
               maxy < this.miny);
14     }
15
16     static BoundingBox union(BoundingBox[] bb) {
17         BoundingBox u = bb[0];
18         for (int i = 1; i < bb.length; i++) {
19             if (bb[i].maxx > u.maxx) {
20                 u.maxx = bb[i].maxx;
21             }
22             if (bb[i].maxy > u.maxy) {
23                 u.maxy = bb[i].maxy;
24             }
25             if (bb[i].minx < u.minx) {
26                 u.minx = bb[i].minx;
27             }
28             if (bb[i].miny < u.miny) {
29                 u.miny = bb[i].miny;
30             }
31         }
32         return u;
33     }
34 }
35
36 class SourceImage {
37
38     int width;
39     int height;
40     BoundingBox bb;
41     BufferedImage image;
42     String name;
43 }
44
45 class TileAddress {
46
47     long row;
48     long column;
49     int scale;
50
51     BoundingBox getBoundingBox() {
52         double dp = 360.0 / (Math.pow(2, scale) * TILE_SIZE);
53         double miny = (row * TILE_SIZE * dp) - 90.0;
54         double maxy = ((row + 1) * TILE_SIZE * dp) - 90.0;
55         double minx = (column * TILE_SIZE * dp) - 180.0;
56         double maxx = ((column + 1) * TILE_SIZE * dp) - 180.0;
57         BoundingBox bb = new BoundingBox(minx, miny, maxx, maxy);
58         return bb;
59     }
60 }
61
62 abstract class TileOutputStream {
63
64     abstract void writeTile(TileAddress address, BufferedImage image);
65
```

```
66  }
67
68  abstract class MemoryImageCache {
69
70      abstract BufferedImage getSourceImageData(String name);
71
72      abstract void putSourceImageData(String name, BufferedImage data);
73  }
```

Listing 5.6 Simple tile creation.

```
1   public void createTiles(SourceImage[] sourceImages, TileOutputStream
        tileOutputStream, int baseLevel, MemoryImageCache cache) {
2
3       //Determine the geographic bounds of the tile set.
4       //This can be based on the bounds of the source images.
5       BoundingBox[] sourceImageBounds = new BoundingBox[sourceImages.length];
6       for (int i = 0; i < sourceImageBounds.length; i++) {
7           sourceImageBounds[i] = sourceImages[i].bb;
8       }
9       BoundingBox tileSetBounds = BoundingBox.union(sourceImageBounds);
10      //Determine the bounds of the tile set in tile coordinates.
11      long mincol = (long) Math.floor((tileSetBounds.minx + 180.0) / (360.0 /
            Math.pow(2.0, (double) baseLevel)));
12      long maxcol = (long) Math.floor((tileSetBounds.maxx + 180.0) / (360.0 /
            Math.pow(2.0, (double) baseLevel)));
13      long minrow = (long) Math.floor((tileSetBounds.miny + 90.0) / (180.0 / Math
            .pow(2.0, (double) baseLevel - 1)));
14      long maxrow = (long) Math.floor((tileSetBounds.maxy + 90.0) / (180.0 / Math
            .pow(2.0, (double) baseLevel - 1)));
15
16      //Iterate over the tile set coordinates.
17      for (long c = mincol; c <= maxcol; c++) {
18          for (long r = minrow; r <= maxrow; r++) {
19              TileAddress address = new TileAddress(r, c, baseLevel);
20              //Compute the geographic bounds of the specific tile.
21              BoundingBox tileBounds = address.getBoundingBox();
22              //Iterate over the source images.
23              BufferedImage tileImage = new BufferedImage(TILE_SIZE, TILE_SIZE,
                    BufferedImage.TYPE_INT_ARGB);
24              for (int i = 0; i < sourceImages.length; i++) {
25                  //Determine if the specific source image intersects the tile
                        being created.
26                  if (sourceImages[i].bb.intersects(tileBounds.minx, tileBounds.
                        miny, tileBounds.maxx, tileBounds.maxy)) {
27                      //Check the cache for the source image.
28                      BufferedImage bi = cache.getSourceImageData(sourceImages[i
                            ].name);
29                      if (bi == null) {
30                          //If it is not in the cache load it from disk and save
                                in the cache.
31                          bi = readImage(sourceImages[i].name);
32                          cache.putSourceImageData(sourceImages[i].name, bi);
33                      }
34                      //Extract the required image data from the source image and
                            store it in the tiled image.
35                      drawImageToImage(bi, sourceImages[i].bb, tileImage,
                            tileBounds);
36                  }
37              }
38              //Save the completed tiled image to the tile storage mechanism.
39              tileOutputStream.writeTile(address, tileImage);
40          }
41      }
42  }
```

Listing 5.7 Scaled tile creation.

```
1   public void createScaledTile(TileInputStream tileInputStream, TileOutputStream
         tileOutputStream, int baseLevel, long minCol, long maxCol,
2           long minRow, long maxRow) {
3       //For each level from base level − 1 to 1, do the following.
4       for (int level = baseLevel − 1; level <= 1; level−−) {
5           //Determine the bounds of the current tile level in tile coordinates.
6           //ratio will be used to reduce the original tile set bounding
                 coordinates to those applicable for each successive level.
7           int ratio = (int) Math.pow(2, baseLevel − level);
8           long curMinCol = (long) Math.floor(minCol / ratio);
9           long curMaxCol = (long) Math.floor(maxCol / ratio);
10          long curMinRow = (long) Math.floor(minRow / ratio);
11          long curMaxRow = (long) Math.floor(maxRow / ratio);
12          //Iterate over the tile set coordinates.
13          for (long c = curMinCol; c <= curMaxCol; c++) {
14              for (long r = curMinRow; r <= curMaxRow; r++) {
15                  //For each tile, do the following:
16                  TileAddress address = new TileAddress(r, c, level);
17                  //Determine the FOUR tiles from the higher level that
                         contribute to the current tile.
18                  TileAddress tile00 = new TileAddress(r * 2, c * 2, level + 1);
19                  TileAddress tile01 = new TileAddress(r * 2, c * 2, level + 1);
20                  TileAddress tile10 = new TileAddress(r * 2, c * 2, level + 1);
21                  TileAddress tile11 = new TileAddress(r * 2, c * 2, level + 1);
22                  //Retrieve the four tile images, or as many as exist.
23                  BufferedImage image00 = tileInputStream.getTile(tile00);
24                  BufferedImage image01 = tileInputStream.getTile(tile01);
25                  BufferedImage image10 = tileInputStream.getTile(tile10);
26                  BufferedImage image11 = tileInputStream.getTile(tile11);
27                  //Combine the four tile images into a single, scaled−down image

28                  BufferedImage tileImage = new BufferedImage(TILE_SIZE,
                         TILE_SIZE, BufferedImage.TYPE_INT_ARGB);
29                  Graphics2D g = (Graphics2D) tileImage.getGraphics();
30                  g.setRenderingHint(RenderingHints.KEY_INTERPOLATION,
                         RenderingHints.VALUE_INTERPOLATION_BILINEAR);
31                  boolean hadImage = false;
32                  if ((image00 != null)) {
33                      g.drawImage(image00, 0, 0, Constants.TILE_SIZE_HALF, Constants
                             .TILE_SIZE_HALF, Constants.TILE_SIZE, 0, 0, Constants.
                             TILE_SIZE,
34                              Constants.TILE_SIZE, null);
35                      hadImage = true;
36                  }
37                  if ((image10 != null)) {
38                      g.drawImage(image10, Constants.TILE_SIZE_HALF, Constants.
                             TILE_SIZE_HALF, Constants.TILE_SIZE, Constants.
                             TILE_SIZE, 0, 0,
39                              Constants.TILE_SIZE, Constants.TILE_SIZE, null);
40                      hadImage = true;
41                  }
42                  if ((image01 != null)) {
43                      g.drawImage(image01, 0, 0, Constants.TILE_SIZE_HALF,
                             Constants.TILE_SIZE_HALF, 0, 0, Constants.TILE_SIZE,
44                              Constants.TILE_SIZE, null);
45                      hadImage = true;
46                  }
47                  if ((image11 != null)) {
48                      g.drawImage(image11, Constants.TILE_SIZE_HALF, 0, Constants
                             .TILE_SIZE, Constants.TILE_SIZE_HALF, 0, 0, Constants.
                             TILE_SIZE,
49                              Constants.TILE_SIZE, null);
50                      hadImage = true;
51                  }
52                  //save the completed tiled image to the tile storage mechanism.
53                  if (hadImage) {
```

```
54              tileOutputStream . writeTile ( address ,  tileImage ) ;
55          }
56      }
57    }
58   }
59 }
```

Chapter 6
Optimization of Tile Creation

The algorithms for creating tile sets presented in the previous chapter represent basic approaches. There are many possible optimizations to make the process more efficient. Some geospatial image sets are small enough that these optimizations are not needed. However, very large image sets will almost always require optimization to make their computation a tractable problem. In this chapter we will present algorithms for the following tiling optimizations:

- Caching tile sets in memory to improve performance
- Partial reading of source images to conserve memory
- Multi-threading of tile creation algorithms
- Tile creation algorithms for distributed computing
- Partial updating of existing tiled image sets

Each of these techniques should be thoroughly considered by those developing a tile creation system. They have been reduced to practice and are essential improvements in an otherwise resource-inefficient process.

6.1 Caching Tile Sets in Memory to Improve Performance

Because reading from and writing to disk are often the most time consuming steps in the tile creation process, we will present an optimized algorithm that minimizes both of these. In the previous chapter we presented two approaches to tile creation: pull-based and push-based. Push-based had the advantage of having to read source images only one time. Pull-based allowed us to write tiled images only one time. We decided to use the pull-based system because, with the addition of a source image cache, we could reduce some of the re-reading of source images. We are still left with the problem of having to re-read the tiled images as we create the lower resolution zoom levels.

However, if our computer system has enough memory to hold all (or a significant subset) or our tiled images, we can use a very efficient push-based approach. We

J.T. Sample and E. Ioup, *Tile-Based Geospatial Information Systems:*
Principles and Practices, DOI 10.1007/978-1-4419-7631-4_6,
© Springer Science+Business Media, LLC 2010

can loop through the source images, read each one once and only once, and apply the data from the source images to our cached tiled images. Only when we have completed looping through the source images will we write our tiled images to memory. Furthermore, since our tiled images remain in memory, there is no need to re-read them when creating the lower resolution levels.

In practice, few systems will have enough RAM to hold a complete tile set, uncompressed, in memory. Therefore, we will need some logical scheme to sub-divide our tile sets into manageable pieces. Recall an earlier example of a high-resolution aerial imagery collection of the world's 50 largest cities. This dataset has a logical separation of tiles built into its structure. We could separately process, in memory, the tiles for each city and merge the results later. For source image sets without logical groupings, we would have to develop some method for geographically partitioning the tile sets. In the next chapter, we will discuss in some detail a general method for solving this problem. For the purposes of the current section assume that such a system is in place.

The algorithm for push-based tile creation with in memory tile cache is as follows:

1. Choose the base level for the tile set.
2. Determine the geographic bounds of the tile set. (This can be based on the bounds of the source images.)
3. Determine the bounds of the tile set in tile coordinates.
4. Initialize the tile cache.
5. Iterate over the source images. For each source image, do the following:

    a. Compute the bounds of the source image in tile coordinates.
    b. Read the source image into memory.
    c. Iterate over the tile set coordinates. For each tile do the following:
        i. Compute the geographic bounds of the tile.
        ii. Check the cache for the tile image. If it is not in the cache, create an empty image and put it in the cache.
        iii. Extract the required image data from the source image and store it in the tiled image.

6. For each level from (base_level - 1) to 1, do the following.

    a. Determine the bounds of the current tile level in tile coordinates.
    b. Iterate over the tile set coordinates. For each tile, do the following:
        i. Determine the four tiles from the higher level that contribute to the current tile.
        ii. Retrieve the four tile images from the cache or as many as exist.
        iii. Combine the four tile images into a single, scaled-down image.
        iv. Save the completed tiled image to the tile cache

7. Finalize the tile cache and store the images on disk

Before presenting the computer code for executing these steps, we will define the data types in Listing 6.1. TileCache represents the mechanism for holding tiled

Listing 6.1 TileCache class.

```
1   abstract class TileCache {
2
3       public abstract BufferedImage getTile(TileAddress ta);
4
5       public abstract void putTile(TileAddress ta, BufferedImage image);
6
7   }
```

images in memory. Additionally, we will make use of the data types defined in the previous chapter. Listing 6.6 shows the algorithm for creating tiles with a memory cache.

6.2 Partial Reading of Source Images

Each of the previously defined algorithms for tile creation assumed that source images can be read and held completely in memory. In some cases this is either not possible or not practical. Some image formats, like MrSID or JPEG2000, support very large images. It is not unusual to encounter images that are several gigabytes compressed.

Uncompressed versions could be 10 times the original size. Even if sufficient memory exists to hold the entire image, we may want to only process a part of the image. Therefore, we need to examine techniques for partial reading of images. Five logical methods for reading images are as follows:

- Whole Image: Only allows users to read entire images in one step.
- Scanlines: Allows users to read one or more scanlines in one step. This is the most common method for low-level access to image pixels.
- Tiles: Allows users to read tiled subsections of images. This is usually only available with image formats that natively store images in subdivided tiles. The reader should note that the concept of "tile" in this context is slightly different from how we have used it throughout the book. In this context, tiles represent rectangular blocks of an image. They are not full images by themselves.
- Random Areas: Allows users to read user-defined rectangular areas of images.

The ability to read only a part of an image is dependent on both the file format used to store the image and the software library used to decode the image. The process is straightforward for uncompressed images. However, it may be impossible with compressed image formats. Java and Python support reading a variety of image formats. However, they do not allow partial reading of images. In general, the most flexible methods for reading images can be found in their C/C++ reference implementations. LIBJPEG, LIBPNG, and LIBTIFF are all open source libraries for reading JPEG, PNG, and TIFF images, respectively. Both LIBJPEG and LIBPNG support scanline based reading. LIBTIFF supports scanline and tile-based reading

depending on how the image was stored. The LizardTech GeoExpress Software
Development Kit (SDK) supports random area reading of MrSID and JPEG2000
images.

Two things are needed to integrate partial reading into our existing tile creation
algorithms. First, we need a method for reading random areas out of our image. This
can be done by directly reading the random areas where supported or by adapting
scanline or tile-based reading to provide random areas. Second, we need to adapt
our tile creation algorithm to account for a partial image instead of the full image.

6.2.1 Reading Random Areas from Source Images

We define a random area as a rectangular region within an image. It is defined by
the coordinates for the origin: x and y, and a width and height (See Figure 6.1).
Before we can extract the image data from the random area we have to determine
the geographic bounds of the intersection between our source image and our tiled
image. We then have to convert the geographic bounds of the intersection area into
source image coordinates.

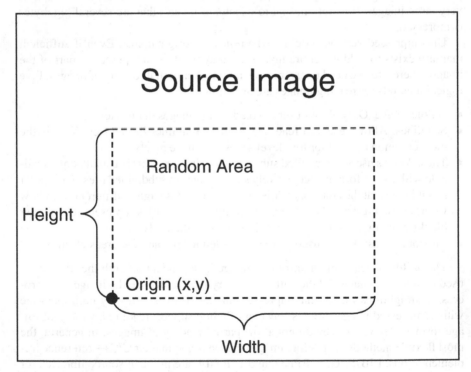

Fig. 6.1 Random area from a source image.

Listing 6.2 Compute the intersection of two bounding rectangles.

```
1  public BoundingBox getIntersection(BoundingBox bb1, BoundingBox bb2) {
2    if (!bb1.intersects(bb2.minx, bb2.miny, bb2.maxx, bb2.maxy)) {
3      return null;
4    }
5    double minx = Math.max(bb1.minx, bb2.minx);
6    double miny = Math.max(bb1.miny, bb2.miny);
7    double maxx = Math.min(bb1.maxx, bb2.maxx);
8    double maxy = Math.min(bb1.maxy, bb2.maxy);
9    BoundingBox out = new BoundingBox(minx, miny, maxx, maxy);
10   return out;
11 }
```

Listing 6.3 Convert geographic bounds to image bounds.

```
1  public Rectangle convertCoordinates(BoundingBox imageBounds, BoundingBox
     subImageBounds, int imageWidth, int imageHeight) {
2
3    int x = (int) Math.round((imageBounds.minx - subImageBounds.minx) / (
       imageBounds.maxx - imageBounds.minx) * imageWidth);
4    int y = imageHeight - (int) Math.round((imageBounds.miny - subImageBounds.
       miny) / (imageBounds.maxy - imageBounds.miny) * imageHeight) - 1;
5    int width = (int) Math.round((subImageBounds.maxx - subImageBounds.minx) /
       (imageBounds.maxx - imageBounds.minx) * imageWidth);
6    int height = (int) Math.round((subImageBounds.maxy - subImageBounds.miny) /
       (imageBounds.maxy - imageBounds.miny) * imageHeight);
7    Rectangle r = new Rectangle(x, y, width, height);
8    return r;
9  }
10
11 class Rectangle {
12   public Rectangle(int x, int y, int width, int height) {
13     this.x = x;
14     this.y = y;
15     this.width = width;
16     this.height = height;
17   }
18
19   int x;
20   int y;
21   int width;
22   int height;
23 }
```

The algorithms shown in Listings 6.2 and 6.3 compute the intersection of two bounding rectangles and convert that intersection from geographic coordinates to image coordinates.

The result of Listing 6.3 is a rectangle in image coordinates. These coordinates can be used to extract a region of pixels from a source image. The algorithms in Listings 6.7 and 6.8 demonstrate how to extract a partial image region with either scanline or tile-based access to a tiled image. Because we are demonstrating low-level access to image data, we will not use a memory image object like Java's Buffered-Image. To represent in-memory images, we will use a simple array of "byte" values that represent RGB pixel values packed in 3 byte triplets and stored in row-major order. Also, the reader should note that the algorithms presented in the following section are written for maximum clarity, not necessarily efficiency. For example, we use "for" loops to copy blocks of bytes while many programming languages have built-in functions that perform this step much faster.

Listing 6.7 presents Java code for reading an image region with scanline based access. The steps for scanline reading of an image region are as follows:

1. Skip or seek to the first scanline needed.
2. For each scanline needed, do the following:

   a. Read the entire scanline into a temporary buffer.
   b. Copy the required subsection of the scanline into the final image buffer.

3. Return the final image buffer.

The algorithm uses the class "ImagePointer" to represent a handle on a file or input stream with encoded pixel data. The algorithm assumes the following functions are available:

- ReadScanline: This function decodes a scanline of pixel data and copies it to the provided buffer. After completion, it positions the image pointer for reading the next available scanline.
- SkipScanlines: This function skips scanlines in the image and positions the image point for reading at the next available scanline. Some image decoding implementations allow random access to scanlines, while others will have to decode the scanlines that are skipped. This difference may affect performance and should be considered by developers.

Listing 6.8 presents Java code for reading partial image regions with tile-based image access. The algorithm assumes that our decoding implementation allows random access to image tiles; this is modeled after LIBTIFF's access routines that do allow random access to image tiles for images that are stored in a certain way.

The steps for tile-based reading of an image region are as follows:

1. Determine the range of tiles that will need to be read.
2. Construct a temporary buffer with sufficient size to hold all of the needed tiles.
3. Iterate over the tiles, in row-major order. For each tile needed, do the following:

   a. Position the image pointer to read at the needed tile.
   b. Read the tile into the temporary buffer.

4. Trim the temporary buffer to match the desired region.

The algorithm also uses the class "ImagePointer" to represent a handle on a file or input stream with encoded pixel data. The algorithm assumes that the following functions are available:

- SeekToTile: This function positions the image pointer to read the indicated tile.
- ReadTile: This function reads pixels from the current tile into the provided buffer.

6.2.2 Tile Creation with Partial Source Image Reading

Listing 6.9 shows our previous algorithm for creating tiles in an adapted form to handle partial reading of source images. The steps in this algorithm are as follows:

1. Chose the base level for the tile set.
2. Determine the geographic bounds of the tile set. (This can be based on the bounds of the source images.)
3. Determine the bounds of the tile set in tile coordinates.
4. Initialize the tile storage mechanism.
5. Iterate over the tile set coordinates. For each tile, do the following:

   a. Compute the geographic bounds of the specific tile.
   b. Iterate over the source images. For each source image do the following:
      i. Determine if the specific source image intersects the tile being created.
      ii. If the source image and tile intersect,
         A. Determine the intersection of tile and source image.
         B. Convert the intersection from geographic to image coordinates.
         C. Read the partial image data.
         D. Convert the partial image data to a BufferedImage.
         E. Draw the converted pixels to the tile image.
   c. Save the completed tiled image to the tile storage mechanism.

6. Finalize the tile storage mechanism.

Within Listing 6.9, the abstract method "ReadPartialImage" is meant as a place-holder for the implementation specific techniques presented in the previous section. The abstract method "ConvertBytes" simply moves the pixel data from the byte array into a Java BufferedImage. Listing 6.9 is a modified form of pull-based tile creation, which should be used if all image tiles cannot be held in memory at the same time. Push-based methods are still preferred if all tiles can be held in memory.

6.3 Tile Creation with Parallel Computing

Parallel computing is the use of multiple computing resources at the same time to execute a given task. This can be realized by a using a group of computer systems or by using a single computer system with multiple CPUs. In most cases, the tile creation process must be parallelized to operate efficiently. The next two sections present techniques for parallelization of the tile creation process from two very different perspectives.

Listing 6.4 Synchronized drawing.

```
1  synchronized (tileImage) {
2       drawImageToImage(bi, currentBounds, tileImage, tileBounds);
3  }
```

6.3.1 Multi-Threading of Tile Creation Algorithms

Multi-threading is a programming technique that, if supported by the underlying operating system and computer hardware, allows multiple tasks to execute at the same time within the same process. A process is an instance of computer program that is being executed. The requirement that the multiple threads of execution exist within the same process is a critical constraint. This allows the multiple threads to share memory with each other. For a detailed tutorial on threading models and usage, the reader is encouraged to see [2].

Multi-threading has three common uses. First, multi-threading is used to manage blocking input/output (I/O). Reading and writing from disk or network is relatively slow, and multi-threading allows the program to perform other tasks while waiting for I/O. A second use is to allow systems with multiple CPUs to process multiple tasks at the same time. The final common use of multi-threading is to make programs with graphical user interfaces more responsive. Multi-threading is useful for this because one thread can be dedicated to updating the graphical interface, while others perform the program's work. The first two of these uses, managing blocking I/O and processing multiple tasks, will be very useful in the tile creation process.

Our first algorithm is derived from our previous push-based tile creation algorithms but adds multi-threading to reduce waiting for I/O. In this algorithm, we have two threads: a reader thread and a tiler thread. The reader thread reads a source image into memory and waits for it to be retrieved by the tiler thread. The tiler thread retrieves decoded source images from the reader thread and creates tiles from it. When the tiler thread takes an image from the reader thread, the reader thread decodes another image and waits for it to be taken. Java code for this process is provided by Listing 6.10. This algorithm should result in a performance improvement, even on systems with just one CPU.

If our processing system has multiple CPUs, and current commodity systems can have up to 48, we can use more than one thread to perform the tiling. This requires two adjustments to the previous algorithm. First, we must create and start more than one tiler thread. Second, we need to make sure that multiple tiling threads are not accessing and writing to the same tile at the same time. To accomplish this, we will change how we call the method for drawing from source image to buffered images. We can wrap the call to the "drawImageToImage" method in a synchronized block that is synchronized on the target BufferedImage, see Listing 6.4. We need to synchronize on only tileImage because that is the only thing getting changed by the various threads. Listing 6.5 shows the method for starting and controlling multiple tiling threads.

Listing 6.5 Controlling multiple tile creation threads.

```
1    public void createTilesMultipleThreads(TileCache cache, SourceImage[]
         sourceImages, int baseLevel, int numberOfThreads) {
2        ReaderThread reader = new ReaderThread(sourceImages);
3        reader.start();
4        TilerThread[] tilerThreads = new TilerThread[numberOfThreads];
5        for (int i = 0; i < tilerThreads.length; i++) {
6            TilerThread tiler = new TilerThread(cache, baseLevel, reader);
7            tiler.start();
8            tilerThreads[i] = tiler;
9        }
10       for (int i = 0; i < tilerThreads.length; i++) {
11           try {
12               tilerThreads[i].join();
13           } catch (InterruptedException e) {
14               e.printStackTrace();
15           }
16       }
17   }
```

It is common to use a number of threads equal to the number of CPUs available. However, in many cases the optimum number of threads can be larger or smaller. This depends on the I/O bandwidth, speed of processors, amount of memory, and other computer specific parameters. Only through trial and error can a developer determine the optimal number of threads.

6.3.2 Tile Creation for Distributed Computing

In the previous section, our multiple lines of execution had the advantage of shared memory to communicate and exchange data. This is not the case for distributed computing since we are spreading our processing across multiple systems called compute nodes. Groups of compute nodes are often called clusters. With distributed computing, communication between nodes is done via a network. In some cases, clusters are connected with dedicated high speed networks like InfiniBand or 10 Gigabit Ethernet. In other cases, compute nodes may be spread out geographically and connected via the Internet. Clusters vary greatly in composition and use. Some clusters fill the traditional role of supercomputers, while others are used to provide services to the public. Some clusters are specially configured groups of identical computer systems while others are ad hoc groupings. Others are made of virtualized systems, dynamically allocated to meet on-demand needs. For the purposes of tile creation, the issues to be considered are nearly the same irrespective of the composition of the cluster.

The two primary tasks related to tile creation using computational clusters are creating a system for breaking the tile creation process into smaller, independent tasks and choosing a software framework for developing the solution. The exact physical configuration of a cluster is less important than these two issues.

The problem of dividing the tile creation process into smaller, independent tasks has already been introduced in Section 6.1. In that section we discussed the requirement for smaller tile creating tasks so that all the tiles could be held in memory. In the context of distributed computing, we have to subdivide our tasks so that we can distribute the source images in smaller collections to individual compute nodes and then collect the created tiles from each node. In the next chapter, we will present a tile storage solution that ties all of these requirements together and presents a general solution for sub-dividing tile creation tasks.

Tile creation is typically an I/O bound problem. As discussed, reading and writing to disk or network is far slower than computing the content of tiles. Tiled image calculations are relatively simple, linear pixel transformations. Given this property, we must minimize data movement to make distributed tile creation a beneficial technique.

In the next sections, we will discuss several software frameworks for distributed computing. In the context of distributed computing, the framework is a software application programming interface (API) that facilitates sharing of data and managing control flow of parallel programs. The chosen software framework usually drives the logical configuration of the computational process. We will introduce the basic concept of each framework and discuss how its properties relate to the tile creation process.

6.3.2.1 MPI

A very common cluster framework is called MPI (Message Passing Interface). MPI is a language independent communications protocol for parallel computing. MPI is just a specification. To be used, a developer must select a concrete implementation of the specification. Fortunately there are several implementations, both open source and commercial. It is most commonly used with the C and FORTRAN languages, although bindings exist for other languages like Java, Python, and the Matlab environment.

MPI provides low-level mechanisms for moving data between nodes, control of execution, and synchronization between independent processes. MPI implementations are very efficient and are a good choice for parallel applications that require a lot of interaction between nodes, as is common for some types of scientific supercomputing. It is also a good candidate for parallel applications that are primarily CPU bound. That is, those that require extensive computations with little I/O. In contrast, tile creation is often I/O bound, especially when it utilizes multi-threading techniques discussed in the previous section.

A fully functional tile creation system could be created utilizing MPI for node-to-node communication and control. However, the relatively low-level nature of MPI commands provides unneeded functionality and would make development a very tedious process. For this reason, we recommend a higher level framework.

6.3.2.2 MapReduce

MapReduce is a distributed computing model created by Google and designed to allow computing problems to be easily solved in a multi-processing environment, from a single shared-memory machine up to a large cluster of heterogeneous networked computers [1]. Users provide map and reduce functions specific to their problem. The MapReduce framework implementation coordinates the distributed computing environment using these functions. The original Google MapReduce implementation is not available to the public. Hadoop[1] is an open source implementation which is commonly used outside of Google.

The MapReduce model is derived from common functional programming techniques. The map function takes an input record in the form of a key/value pair. The map function then processes that input and creates a new intermediate key/value pair. The reduce function takes one of these intermediate keys as input as well as its associated values. The function then merges these values, usually into a single value. The MapReduce framework distributes input records to the map function and receives its output. It then distributes those intermediate values to the reduce function and receives the merged results. Communication and errors are also managed by the MapReduce framework.

MapReduce is most effectively used when the target problem is computationally bound, the input has a large number of records, or the distributed computing platform is large and complex. The MapReduce framework handles most of the management while a user only need implement the details for a specific task process. The benefits of MapReduce are dependent on the implementation being used. Different MapReduce implementations may provide different I/O capabilities, management capabilities, and error handling capabilities. For example, the Hadoop implementation does not allow random writes within files in its filesystem.

MapReduce may be used as a framework for tiling in a cluster, however, its capabilities are not necessarily well aligned to the task because the tiling process is I/O bound. The source data, large image files, and the output data, large tile files, must be moved to and from the processing systems. The computational cost of processing the imagery is much smaller than the I/O cost. The distributed file system used by the MapReduce framework (HDFS for Hadoop) will incur as much or more I/O cost by positioning files throughout the network. Retrieving data for use elsewhere will incur the same penalties. Additionally, large clusters are not necessary to tile map imagery. Fewer than 20 (potentially fewer than 10) processing nodes need be used to process even the largest imagery datasets in a reasonable amount of time (a few days). Given that the tile creation process is perfectly parallelizable, the complexity of the overall tile processing system is not large enough to support the use of a MapReduce framework.

The MapReduce model works well for a user such as Google because they have a large number of diverse distributed computing problems that may be solved using one framework. They also have a large and geographically diverse computing clus-

---

[1] http://hadoop.apache.org/

ter, which would be difficult to manage without a model such as MapReduce. Tiling is not computationally bound, has a relatively limited number of image inputs, and may be run on smaller cluster systems. Unless a MapReduce framework is useful to other applications in the enterprise, it is not necessary to use for tiling. The cost of installing the MapReduce framework combined with the cost of implementing the tile processing in using the MapReduce framework will not be significantly lower than simply creating an ad hoc clustering system.

6.3.2.3 Ad Hoc Clustering

Ad hoc clustering refers to distributed computing with no specific software framework. Software frameworks often provide useful tools, but they also introduce overhead either at execution time or development time. There are many ways to control program execution and data sharing between networked computers. REXEC or Secure Shell (SSH) can be used to remotely start and control processes on networked systems. Server Message Block (SMB) and Network File System (NFS) can be used to share data across a network through remotely mounted file systems.

Given the data intensive nature of tile creation processes, developers should create distributed systems with very minimal interactions between systems. With this constraint, sophisticated communication and control frameworks should be needed only in cases where truly large numbers of CPUs are being controlled. We typically create multi-terabyte tile sets on a cluster with 64 CPUs in just a few hours of compute time. Those CPUs are controlled with SSH commands and share data via NFS.

6.4 Partial Updating of Existing Tiled Image Sets

In the previous sections on tile creation, we have assumed all tile sets are created from source images in one single and final step. How, then, should we handle cases in which new source images need to be added to an existing tile set? This is a common problem for tile sets based on satellite or aerial imagery. These sensing platforms can image only a small portion of the earth's surface. A complete picture of a sizable area will include source images taken over an extended time period.

The most basic approach to handling updated images is to simply discard the previous tile set and create a new one each time new source images are available. In some cases, this is the best approach. If a majority of the source images have been updated or if the tile set is rather small, it may be just as efficient to start over. However, if the existing tile set is large and the updates are relatively small, starting over would be expensive or even impossible. Consider a very large example tile set that takes two weeks to create. If some source images are updated every week, we would have to start processing a new tile set as soon as the previous one finished. To keep up-to-date, we would always be processing the large tile set, and most of our processing would be redundant. This would be expensive both in our time and computational resources.

A better approach is to integrate updated source images into existing tile sets by altering the contents of only the tiles that are affected by the new source images. Logically, the change in our tile algorithm is very simple. Instead of creating a new empty image for a tile, we simply retrieve the existing tile from storage, update its contents, and store the new image. The main difficulty lies in developing a tile storage system capable of handling updated image files. Another challenge is that we must maintain sufficient source image and tile metadata so we can detect which source images should be added to the tile set. Both of these problems relate directly to tile storage and will be discussed in the next chapter.

Listing 6.6 Push-based tile creation with a memory tile cache.

```
1   public void createCachedTiles(TileCache cache, SourceImage[] sourceImages, int
        baseScale) {
2       //Determine the geographic bounds of the tile set.
3       //This can be based on the bounds of the source images.
4       BoundingBox[] sourceImageBounds = new BoundingBox[sourceImages.length];
5       for (int i = 0; i < sourceImageBounds.length; i++) {
6   sourceImageBounds[i] = sourceImages[i].bb;
7       }
8       BoundingBox tileSetBounds = BoundingBox.union(sourceImageBounds);
9       //Determine the bounds of the tile set in tile coordinates.
10      long tilesetMincol = (long) Math.floor((tileSetBounds.minx + 180.0) /
            (360.0 / Math.pow(2.0, (double) baseScale)));
11      long tilesetMaxcol = (long) Math.floor((tileSetBounds.maxx + 180.0) /
            (360.0 / Math.pow(2.0, (double) baseScale)));
12      long tilesetMinrow = (long) Math.floor((tileSetBounds.miny + 90.0) / (180.0
            / Math.pow(2.0, (double) baseScale - 1)));
13      long tilesetMaxrow = (long) Math.floor((tileSetBounds.maxy + 90.0) / (180.0
            / Math.pow(2.0, (double) baseScale - 1)));
14
15      //Iterate over the source images
16      for (int i = 0; i < sourceImages.length; i++) {
17      BoundingBox currentBounds = sourceImages[i].bb;
18      //Compute the bounds of the source image in tile coordinates
19      long mincol = (long) Math.floor((currentBounds.minx + 180.0) / (360.0 / Math.
            pow(2.0, (double) baseScale)));
20      long maxcol = (long) Math.floor((currentBounds.maxx + 180.0) / (360.0 / Math.
            pow(2.0, (double) baseScale)));
21      long minrow = (long) Math.floor((currentBounds.miny + 90.0) / (180.0 / Math.
            pow(2.0, (double) baseScale - 1)));
22      long maxrow = (long) Math.floor((currentBounds.maxy + 90.0) / (180.0 / Math.
            pow(2.0, (double) baseScale - 1)));
23      //Read the source image into memory
24      BufferedImage bi = readImage(sourceImages[i].name);
25      for (long c = mincol; c <= maxcol; c++) {
26          for (long r = minrow; r <= maxrow; r++) {
27          TileAddress address = new TileAddress(r, c, baseScale);
28          //Compute the geographic bounds of the specific tile.
29          BoundingBox tileBounds = address.getBoundingBox();
30          //Check the TileCache for the tiled image
31          BufferedImage tileImage = cache.getTile(address);
32          if (tileImage == null) {
33              tileImage = new BufferedImage(TILE_SIZE, TILE_SIZE, BufferedImage.
                    TYPE_INT_ARGB);
34              cache.putTile(address, tileImage);
35              //Extract the required image data from the source image and store it in
                    the tiled image.
36              drawImageToImage(bi, sourceImages[i].bb, tileImage, tileBounds);
37              //Note that since tileImage is a pointer to the bufferedimage already
                    in the cache,
38              //we don't have to put it back in after each use.
39          }
40          }
41      }
42      }
43      for (int scale = baseScale - 1; scale <= 1; scale--) {
44      //Determine the bounds of the current tile scale in tile coordinates.
45      //ratio will be used to reduce the original tile set bounding coordinates to
                those applicable for each successive scale.
46      int ratio = (int) Math.pow(2, baseScale - scale);
47      long curMinCol = (long) Math.floor(tilesetMincol / ratio);
48      long curMaxCol = (long) Math.floor(tilesetMaxcol / ratio);
49      long curMinRow = (long) Math.floor(tilesetMinrow / ratio);
50      long curMaxRow = (long) Math.floor(tilesetMaxrow / ratio);
51      //Iterate over the tile set coordinates.
52      for (long c = curMinCol; c <= curMaxCol; c++) {
53          for (long r = curMinRow; r <= curMaxRow; r++) {
```

```
54    //For each tile, do the following:
55    TileAddress address = new TileAddress(r, c, scale);
56    //Determine the FOUR tiles from the higher scale that contribute to the
         current tile.
57    TileAddress tile00 = new TileAddress(r * 2, c * 2, scale + 1);
58    TileAddress tile01 = new TileAddress(r * 2, c * 2, scale + 1);
59    TileAddress tile10 = new TileAddress(r * 2, c * 2, scale + 1);
60    TileAddress tile11 = new TileAddress(r * 2, c * 2, scale + 1);
61    //Retrieve the four tile images, or as many as exist.
62    BufferedImage image00 = cache.getTile(tile00);
63    BufferedImage image01 = cache.getTile(tile01);
64    BufferedImage image10 = cache.getTile(tile10);
65    BufferedImage image11 = cache.getTile(tile11);
66    //Combine the four tile images into a single, scaled-down image.
67    BufferedImage tileImage = new BufferedImage(TILE_SIZE, TILE_SIZE,
         BufferedImage.TYPE_INT_ARGB);
68    Graphics2D g = (Graphics2D) tileImage.getGraphics();
69    g.setRenderingHint(RenderingHints.KEY_INTERPOLATION, RenderingHints.
         VALUE_INTERPOLATION_BILINEAR);
70    boolean hadImage = false;
71    if ((image00 != null)) {
72        g.drawImage(image00, 0, Constants.TILE_SIZE_HALF, Constants.
             TILE_SIZE_HALF, Constants.TILE_SIZE, 0, 0, Constants.TILE_SIZE,
73        Constants.TILE_SIZE, null);
74        hadImage = true;
75    }
76    if ((image10 != null)) {
77        g.drawImage(image10, Constants.TILE_SIZE_HALF, Constants.TILE_SIZE_HALF
             , Constants.TILE_SIZE, Constants.TILE_SIZE, 0, 0,
78        Constants.TILE_SIZE, Constants.TILE_SIZE, null);
79        hadImage = true;
80    }
81    if ((image01 != null)) {
82        g.drawImage(image01, 0, 0, Constants.TILE_SIZE_HALF, Constants.
             TILE_SIZE_HALF, 0, 0, Constants.TILE_SIZE,
83        Constants.TILE_SIZE, null);
84        hadImage = true;
85    }
86    if ((image11 != null)) {
87        g.drawImage(image11, Constants.TILE_SIZE_HALF, 0, Constants.TILE_SIZE,
             Constants.TILE_SIZE_HALF, 0, 0, Constants.TILE_SIZE,
88        Constants.TILE_SIZE, null);
89        hadImage = true;
90    }
91    //save the completed tiled image to the tile storage mechanism.
92    if (hadImage) {
93        cache.putTile(address, tileImage);
94    }
95      }
96  }
97    }
98 }
```

Listing 6.7 Read an image region with scanline based access.

```
 1  abstract void skipScanlines(ImagePointer im, int num);
 2
 3  abstract void readScanline(ImagePointer im, byte[] scanlineBuffer);
 4
 5  byte[] readScanlines(ImagePointer im, int imageWidth, int imageHeight, int x,
        int y, int height, int width) {
 6      byte[] outputImage = new byte[imageWidth * imageHeight * 3];
 7      int startScanline = y − 1;
 8      skipScanlines(im, startScanline);
 9      byte[] tempBuffer = new byte[imageWidth * 3];
10      int imageCounter = 0;
11      int scanlineOffset = x * 3;
12      for (int i = 0; i < height; i++) {
13  readScanline(im, tempBuffer);
14      for (int j = 0; j < (width * 3); j++) {
15          outputImage[imageCounter] = tempBuffer[j + scanlineOffset];
16          imageCounter++;
17      }
18      }
19      return outputImage;
20  }
```

Listing 6.8 Read a partial image region with tile-based image access.

```
 1      abstract void seekToTile(ImagePointer im, int i, int j);
 2
 3      abstract void readTile(ImagePointer im, byte[] tileBuffer);
 4
 5      byte[] readTiles(ImagePointer im, int imageWidth, int imageHeight, int
            tileWidth, int tileHeight, int x, int y, int height, int width) {
 6
 7          //Determine the range of tiles that will need to be read.
 8          double numtiles = (Math.ceil((double) imageWidth / tileWidth)) * (Math.
                ceil((double) imageHeight / tileHeight));
 9
10          int startXTile = (int) Math.floor((double) x / tileWidth);
11          int startYTile = (int) Math.floor((double) y / tileHeight);
12          int endx = x + width − 1;
13          if (endx > imageWidth) {
14              endx = imageWidth;
15          }
16          int endy = y + height − 1;
17          if (endy > imageHeight) {
18              endy = imageHeight;
19          }
20          int endXTile = (int) Math.floor((double) endx / tileWidth);
21          int endYTile = (int) Math.floor((double) endy / tileHeight);
22
23          int tileSizeBytes = tileWidth * tileHeight * 3;
24
25          int numtilesToDecode = (endXTile − startXTile + 1) * (endYTile −
                startYTile + 1);
26
27          //Construct a temporary buffer with sufficient size to hold all of the
                needed tiles.
28          byte[] tempImage = new byte[numtilesToDecode * tileSizeBytes];
29
30          int tempImageRowWidth = (endXTile − startXTile + 1) * 3 * tileWidth;
31
32          byte[] tileBuffer = new byte[tileSizeBytes];
33
34          int startYTileCoord = startYTile * tileHeight;
35          int startXTileCoord = startXTile * tileWidth;
36          int bufferOffset = 0;
37          //Iterate over the tiles, in row−major order.
```

```
38      for (int ty = startYTile; y <= endYTile; y++) {
39          for (int tx = startXTile; x <= endXTile; x++) {
40              //Position the image pointer to read at the needed tile.
41              seekToTile(im, tx, ty);
42              //Read the tile into the temporary buffer.
43              readTile(im, tileBuffer);
44              int bufferStartYTile = (ty - startYTile);
45              int bufferStartXTile = (tx - startXTile);
46              int bufferStartYPixel = bufferStartYTile * tileHeight;
47              for (int m = 0; m < tileHeight; m++) {
48                  int startRow = (bufferStartYPixel + m) * tempImageRowWidth;
49                  int startColumn = bufferStartXTile * 3 * tileWidth;
50                  for (int n = 0; n < tileWidth * 3; n++) {
51                      tempImage[startRow + startColumn + n] = tileBuffer[m *
                            tileWidth * 3 + n];
52                  }
53              }
54          }
55      }
56      //Trim the temporary buffer to match the desired region.
57      int xOffset = x - startXTile * tileWidth;
58      int yOffset = y - startYTile * tileHeight;
59      byte[] outputImage = new byte[imageWidth * imageHeight * 3];
60      int imageCounter = 0;
61      for (int i = 0; i < imageHeight; i++) {
62          int rowOffset = (yOffset + i) * tileWidth;
63          for (int j = 0; j < imageWidth; j++) {
64              int columnOffset = j + yOffset;
65              outputImage[imageCounter] = tempImage[rowOffset + columnOffset
                    ];
66              imageCounter++;
67          }
68      }
69      return outputImage;
70  }
```

Listing 6.9 Tile creation with partial source image reading.

```
1   abstract byte[] readPartialImage(String name, int x, int y, int width, int
        height);
2
3   abstract BufferedImage convertBytes(byte[] pixels);
4
5   public void createTilesWithPartialReading(SourceImage[] sourceImages,
        TileOutputStream tileOutputStream, int baseLevel) {
6
7       //Determine the geographic bounds of the tile set.
8       //This can be based on the bounds of the source images.
9       BoundingBox[] sourceImageBounds = new BoundingBox[sourceImages.length];
10      for (int i = 0; i < sourceImageBounds.length; i++) {
11          sourceImageBounds[i] = sourceImages[i].bb;
12      }
13      BoundingBox tileSetBounds = BoundingBox.union(sourceImageBounds);
14      //Determine the bounds of the tile set in tile coordinates.
15      long mincol = (long) Math.floor((tileSetBounds.minx + 180.0) / (360.0 /
            Math.pow(2.0, (double) baseLevel)));
16      long maxcol = (long) Math.floor((tileSetBounds.maxx + 180.0) / (360.0 /
            Math.pow(2.0, (double) baseLevel)));
17      long minrow = (long) Math.floor((tileSetBounds.miny + 90.0) / (180.0 /
            Math.pow(2.0, (double) baseLevel)));
18      long maxrow = (long) Math.floor((tileSetBounds.maxy + 90.0) / (180.0 /
            Math.pow(2.0, (double) baseLevel - 1)));
19
20      //Iterate over the tile set coordinates.
21      for (long c = mincol; c <= maxcol; c++) {
22          for (long r = minrow; r <= maxrow; r++) {
```

```
23        TileAddress address = new TileAddress(r, c, baseLevel);
24        //Compute the geographic bounds of the specific tile.
25        BoundingBox tileBounds = address.getBoundingBox();
26        //Iterate over the source images.
27        BufferedImage tileImage = new BufferedImage(TILE_SIZE,
              TILE_SIZE, BufferedImage.TYPE_INT_ARGB);
28        for (int i = 0; i < sourceImages.length; i++) {
29            //Determine if the specific source image intersects the
                  tile being created.
30            if (sourceImages[i].bb.intersects(tileBounds.minx,
                  tileBounds.miny, tileBounds.maxx, tileBounds.maxy)) {
31                //Determine intersection of tile and source image
32                BoundingBox partialBB = getIntersection(sourceImages[i
                  ].bb, tileBounds);
33                //Convert geographic coordinates to image coordinates
34                Rectangle rectangle = convertCoordinates(sourceImages[i
                  ].bb, partialBB, sourceImages[i].width,
                  sourceImages[i].height);
35                //Read partial image data
36                byte[] data = readPartialImage(sourceImages[i].name,
                  rectangle.x, rectangle.y, rectangle.width,
                  rectangle.height);
37                //convert the pixel bytes to a BufferedImage
38                BufferedImage bi = convertBytes(data);
39                //Draw the converted pixels to the tile image
40                drawImageToImage(bi, partialBB, tileImage, tileBounds);
41            }
42        }
43        //Save the completed tiled image to the tile storage mechanism.
44        tileOutputStream.writeTile(address, tileImage);
45    }
46   }
47  }
```

Listing 6.10 Tile creation with a reader and tiler threads.

```
1  public void createTilesTwoThreads(TileCache cache, SourceImage[] sourceImages,
       int baseLevel) {
2        ReaderThread reader = new ReaderThread(sourceImages);
3        reader.start();
4        TilerThread tiler = new TilerThread(cache, baseLevel, reader);
5        tiler.start();
6        tiler.join();
7   }
8
9   class TilerThread extends Thread {
10
11      private TileCache cache;
12      private int baseLevel;
13      private ReaderThread reader;
14
15      public TilerThread(TileCache tileCache, int baseLevel, ReaderThread
              reader) {
16          this.cache = tileCache;
17          this.baseLevel = baseLevel;
18          this.reader = reader;
19      }
20
21      public void run() {
22
23          ImageWrapper image = reader.getImage();
24          while (image != null) {
25              BoundingBox currentBounds = image.si.bb;
26              long mincol = (long) Math.floor((currentBounds.minx + 180.0) /
                      (360.0 / Math.pow(2.0, (double) baseLevel)));
```

```
27      long maxcol = (long) Math.floor((currentBounds.maxx + 180.0) /
                      (360.0 / Math.pow(2.0, (double) baseLevel)));
28      long minrow = (long) Math.floor((currentBounds.miny + 90.0) /
                      (180.0 / Math.pow(2.0, (double) baseLevel - 1)));
29      long maxrow = (long) Math.floor((currentBounds.maxy + 90.0) /
                      (180.0 / Math.pow(2.0, (double) baseLevel - 1)));
30      BufferedImage bi = image.bi;
31      for (long c = mincol; c <= maxcol; c++) {
32          for (long r = minrow; r <= maxrow; r++) {
33              TileAddress address = new TileAddress(r, c, baseLevel);
34              BoundingBox tileBounds = address.getBoundingBox();
35              BufferedImage tileImage = cache.getTile(address);
36              if (tileImage == null) {
37                  tileImage = new BufferedImage(TILE_SIZE, TILE_SIZE,
                              BufferedImage.TYPE_INT_ARGB);
38                  cache.putTile(address, tileImage);
39
40              }
41              drawImageToImage(bi, currentBounds, tileImage,
                      tileBounds);
42          }
43      }
44      image = reader.getImage();
45      }
46  }
47  }
48
49  class ReaderThread extends Thread {
50
51      List<SourceImage> images = Collections.synchronizedList(new ArrayList<
                  SourceImage>());
52
53      ImageWrapper currentImage = null;
54
55      public ReaderThread(SourceImage[] images) {
56          for (int i = 0; i < images.length; i++) {
57              this.images.add(images[i]);
58          }
59      }
60
61      public void run() {
62
63          while (images.size() > 0) {
64              if (currentImage == null) {
65                  SourceImage si = images.remove(0);
66                  BufferedImage bi = readImage(si.name);
67                  ImageWrapper iw = new ImageWrapper(si, bi);
68                  currentImage = iw;
69              }
70              try {
71                  Thread.sleep(200);
72              } catch (InterruptedException e) {
73                  e.printStackTrace();
74              }
75          }
76
77      }
78
79      public synchronized ImageWrapper getImage() {
80          ImageWrapper returnVal = null;
81          while (currentImage == null) {
82              if (images.size() == 0) {
83                  return null;
84              }
85              try {
86                  Thread.sleep(400);
87              } catch (InterruptedException e) {
```

```
88              e . printStackTrace ( ) ;
89          }
90      }
91      returnVal = currentImage ;
92      currentImage = null ;
93
94      return returnVal ;
95      }
96   }
97
98   class ImageWrapper {
99
100     BufferedImage  bi ;
101     SourceImage  si ;
102
103     public ImageWrapper ( SourceImage  si ,  BufferedImage  bi ) {
104         super ( ) ;
105         this . si  =  si ;
106         this . bi  =  bi ;
107     }
108
109   }
```

References

1. Dean, J., Ghemawat, S.: Map Reduce: Simplified data processing on large clusters. Communications of the ACM-Association for Computing Machinery-CACM **51**(1), 107–114 (2008)
2. Oaks, S.: Java Threads. O'Reilly (2004)

Chapter 7
Tile Storage

The two previous chapters presented several algorithms for creation of tiled images. Each of those algorithms assumed that some mechanism was in place to support storage and retrieval of tiled images. In this chapter, we will discuss such mechanisms and provide technical guidance on choosing a tile storage system. We will also discuss some advanced topics in tile storage, such as storage of tile metadata and distributed storage of tiles.

7.1 Introduction to Tile Storage

Tiled image layers are divided into levels. Each level is divided into rows and columns. Figure 7.1 shows a tiled layer divided into levels, then columns, and then tiles. The general problem of tile storage is linking a tile's address (Layer, Level, Row, and Column) to a binary block of data. That linking should be quickly generated, retrieved, or altered. The practical problem of tile storage is how to organize the blocks of data into levels, rows, and columns so that the tiled images can be efficiently written to and read from disk.

All tiled images are stored in computer files on disk. Tiles can be stored in a separate file for each image, bundled together into larger files, or in database tables. (Database systems use files like any other computer program, so storing tiles in a database indirectly stores them to file.)

The next three sections provide detailed explanations of alternative methods for storing tiles in files. A fourth section provides performance comparisons between the three methods.

J.T. Sample and E. Ioup, *Tile-Based Geospatial Information Systems:*
Principles and Practices, DOI 10.1007/978-1-4419-7631-4_7,
© Springer Science+Business Media, LLC 2010

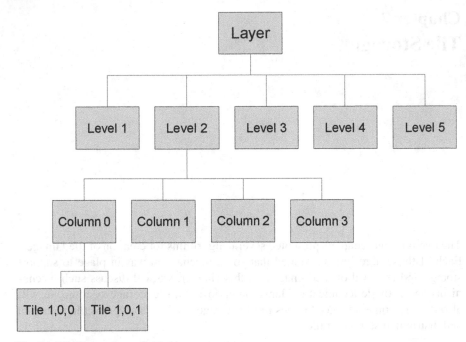

Fig. 7.1 Tiled image layer divided into components.

7.2 Storing Image Tiles as Separate Files

A simple and common method for storing tiled images is to simply store each image in a separate computer file on the computer's file system. Recall from Chapter 5, our tiled images are formatted in standard image formats, like JPEG or PNG. Each of these formats was designed to store an image as a single computer file. Folders (or directories) on the file system can be used to provide structure and organization to the tiled images. For example, we can use a top level folder for the layer, sub-folders within the layer folder for each level, and then subfolders within the level folders for each column. Within the column folders are the individual tiled images for each row in that column. Figure 7.2 shows such an organization.

This type of organization is attractive to developers for several reasons. First, tiles can be addressed directly by simply forming the filename and opening the file. For example, if I want to create a tile, for layer "BlueMarble" at level 7, column 5, and row 4 I can simply create the string "BlueMarble/7/5/4.jpg" and I have the filename for the desired tile. With this method, there is no need for a separate index of tiles. A second benefit is that tiled images can be replaced by newer versions with little impact on the rest of the system.

Finally, and most importantly, building a Web server to host the tiled images in this structure is trivial. Most HTTP servers, including Apache, can, by default, host files directly on the file system accessible by the sub-path. So, to access a tile over the web, I can construct a URL like the following:

`http://www.sometileserver.com/BlueMarble/7/5/4.jpg`

The HTTP server will simply retrieve the image directly from the file system with minimal configuration.

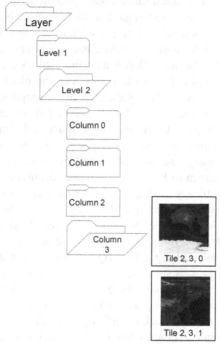

Fig. 7.2 Folder based organization of tiled images.

However, there are several disadvantages of storing tiles in this method. From the perspective of a software developer, file systems can appear to function by magic. A developer simply names the file he wants, and it appears. He can add to it, delete it, or move it. The system magically knows the size and location of the file, the date it was modified, and which users have what permissions on the file.

In reality, file systems are among the most complicated parts of an operating system. Even though a file can be created with a single line of computer code, there are many things going on behind the scenes that enable that file to magically appear. Space on the hard drive has to be located and allocated for the file. Lists of blocks used to store the file have to be written along with the file's metadata. To store this information, file systems have their own meta-storage allocated. The file system's meta-storage has to be accessed for every file that is created or accessed. In everyday use these operations often seem instant because modern operating systems can cache the file system's meta-storage in memory. However, when writing and reading many millions of files the memory cache will fail to hold all the needed information, and the file accesses will take much longer. When the small price of a single file access is added to the creation of each and every tile, this method becomes very inefficient and unsuitable for very large tile sets.

Additionally, many file systems do not index files by name. File lookups involve a linear search within a given directory. This is especially problematic given our structure in which a single column folder could hold thousands of image files.

Files are somewhat wasteful with regards to storage space because files are stored in fixed size blocks. A common block size is 4096 bytes, so a file will be broken up into pieces of this size. Files almost always consume an uneven number of blocks. For example, a 10000 byte file will consume three blocks, and a total of 12288 bytes. The average wasted data per file is one half the block size. If the average size of a tiled image file is 50,000 bytes, then the average wasted space is 2048 bytes. Therefore we are wasting around 4% of our storage space with this approach. Four% would be a small price to pay in storage space if this approach yielded significant performance improvements. However, since this approach will likely yield significant performance degradations, the wasted space adds insult to injury.

In many cases tile sets must be copied from one location to another. Perhaps the system that created the tiles is not the same one that will serve them to users over a network, or perhaps multiple systems will be used to serve the same set of tiles. In these cases copies of the entire tile set must be created. To create a copy of the tile set with this storage method requires a separate file access and file write for each tile. This process can take as long as the original tile creation step.

In general, storing tiles as separate image files is a horribly inefficient use of the computer's resources. However, there are a few scenarios in which this is a good approach. First, when dealing with very small tile sets, those with only a few thousand tiled images, this approach is perfectly valid. A more complicated solution would be a waste of time. Second, when the inherent properties of the file system are actually needed, this approach might be useful. For example, a developer might need full use of permissions on each and every tiled image. If the tiles are updated very frequently, and the older tiles can be discarded, this approach might be valid. File systems have sophisticated methods of recapturing used storage space that is no longer needed. Frequent changes to tiles would necessitate this capability.

There is one final scenario in which storing tiles as separate image files makes sense. The File System in Userspace (FUSE) API[1] allows developers to create custom file systems that mimic the properties of a file system on the front end, but store the actual file data with a custom method defined by the developer. A FUSE file system implementation could be created that would allow tiles to be written by software as separate files. On the back end, the tiled images would be stored in an efficient manner that eliminates much of the overhead associated with full featured file systems. This FUSE implementation would also integrate with the HTTP server used to distribute the tiled images. This approach would allow tile system developers

---

[1] http://fuse.sourceforge.net/

to use a variety of existing, open-source tile creation and distribution tools on very large tile sets.

7.3 Database-Based Tile Storage

A second approach to storing tiled images is to store the images within a relational database management system (DBMS) as binary large objects (BLOB). Most modern database systems allow arbitrary size binary arrays to be stored along side structured columns. Using this approach, we can build a "tiles" table with a column for the image data and other columns for the address components of the tile: level, row, and column. This approach is slightly more complicated than simply storing the data in files. However, since modern database systems use sophisticated techniques for paging of storage this approach might be more efficient. Additionally, we can create indices on the address columns, which could reduce search time.

A disadvantage of this approach is that database systems can be costly in terms of expense, setup, configuration, and maintenance. Like the file system approach, this approach brings a lot of unneeded features that may introduce overhead into the system. Database systems are designed to operate on highly structured data, such as small numeric and character fields. A tile storage system has little need for queries on structured data. Databases also excel at revision control which is unlikely to be needed for a tile system.

As will become apparent in the Comparative Performance section, databases are unlikely to be widely used for storage of tiles. However, there are a couple of scenarios in which they may apply. First, some commercial Web hosting systems provide users with read/write access to a database but not to the file system. If we were forced to use this type of system, we would have to store our tiles in a database. Secondly, if our tile application required sophisticated query functionality we might need a database. For example, if our tiled images also came with extensive metadata like dates, places, names, and keywords that need to be queried for tile retrieval, a database would be useful. A database/file hybrid approach is also a possibility. In this case, the tiles metadata and addresses would be stored in a database, and the image data stored in large flat files.

7.4 Custom File Formats

Another approach to storing tiled images is to use a custom designed file format. In this case many tiled images are packed together in a single file instead of in multiple files. This approach necessitates development of an organizational system to keep up with the locations of the tiled images in the single file. It also requires a custom index that allows lookup of tile positions within the large files. This method can offer vastly improved performance, since the inefficiencies of the underlying file system are mitigated. Another benefit is that the large custom files can more easily be copied from one location to another than many millions of smaller files.

A disadvantage of this system is that, if tiled images change frequently, the custom files may become fragmented. That is, they are littered with out-of-date tiled images that need to be cleaned up. Another disadvantage is that the tiled images cannot be directly accessed by an HTTP server. The server will need a custom module to read the custom formatted files.

In the next chapter we will present two methods for storing images in custom file formats. We will explain the tiled image organization system as well as some high performance indexing schemes.

7.5 Comparative Performance

The three previous sections have explained three alternative methods for storing tiled images. In each of those sections we presented some conceptual and practical advantages and disadvantages of each method. In this section we will use some test programs to show the differences in performance.

Benchmarking file writing and reading is very challenging. Modern operating systems perform a lot of caching that can interfere with the results. The best way to measure performance is to create benchmarks that are very close to real-world tasks and run those many times. In this fashion, you can replicate a realistic user environment and average out anomalous results. Before each test we will clear the file system's cache by executing the following Linux command as superuser:

```
echo 3 > /proc/sys/vm/drop_caches
```

This will help ensure each test is performed in a similar environment. The hardware and software configuration for these tests is the same for all tests and is listed in Table 7.1.

Operating System	Debian 5
Java Virtual Machine	1.6.0_15 (64 bit)
DBMS	Postgres 8.4, default configuration
Processors	2 2.0Ghz AMD Opteron
RAM Size	16GB DDR2 776Mhz
Hard Drive Specification	Dell MD1000 with 15 1TB SAS drives
File System	XFS

Table 7.1 Test configuration.

7.5.1 Writing Tests

This first set of tests will examine writing tiled images. We will write a large number of tile-sized pieces of memory to disk in three different ways and compare the results. In each of the writing tests, we will write tile-sized pieces for each tile in zoom levels 5 through 11. Zoom levels 5 though 11 have 512; 2,048; 8,192; 32,768; 131,072; 524,288; and 2,097,152 tiles, respectively. Each piece of data will be 50,000 bytes in length. The data we write will be simple arrays of random or zero data. We are concerned only with testing the different types of I/O, so the actual contents of the files are not important. We will run each test 30 times to get average performance numbers.

To represent the three methods, we have written three simple implementations. The first implementation writes each tile to a separate file. The second implementation writes all the tiles into a single file for each zoom level and includes an index of tile locations. The third implementation writes all the tiles into a single database table for each zoom level. Each test writes the data to new files and not over existing files.

Listing 7.1 shows the three implementations. In the section "WriteTilesSingleFile" we reference the classes IndexedTileOutputStream and IndexedTileInputStream. These classes are part of the first tile storage implementation discussed in the next chapter and their code is presented there. Table 7.2 shows the results from running the write tests 30 times each. The mean times are in seconds. ¿From this

Level	Number of Tiles	Single File per Tile		Single File per Level		Database Table per Level	
		Mean	StdDev	Mean	StdDev	Mean	StdDev
5	512	0.1049	0.027	0.086	0.022	0.683	0.033
6	2,048	0.8477	0.075	0.257	0.029	2.654	0.198
7	8,192	3.5807	1.623	1.090	0.115	10.540	0.509
8	32,768	14.2025	1.857	3.795	0.187	42.145	1.140
9	131,072	56.7045	2.567	21.532	0.265	167.979	3.964
10	524,288	244.9717	3.862	91.684	0.695	673.950	12.783
11	2,097,152	999.9249	27.582	383.365	2.762	2767.647	67.018

Table 7.2 Mean times in seconds and standard deviations from 30 trials of write tests.

table we can see that writing multiple tiles to a single large file yields the best performance. Writing each tile to a separate file takes 2 to 3 times the amount of time. Writing tiles to a DBMS takes 5 to 10 times the amount of time. Figure 7.3 plots the results in terms of average write per tile. The write times for each level are fairly consistent.

Many DBMS systems support bulk imports of data. It would be possible to write tiles out using the fast single file method and then import the data into the database. We have not benchmarked this procedure. Though it would offer some improvement in write performance, it would still be slower than simply writing to the single file. We will see in the next section that reading from the database is also significantly slower.

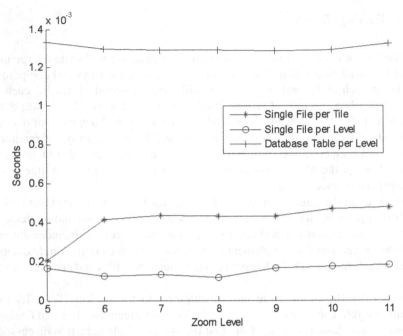

Fig. 7.3 Plot of average write times per tile.

7.5.2 *Reading Tests*

For the reading tests, we will use the tiles written in the previous step. The first test will mimic random access of tiles stored on disk, and the second test will mimic random access of tiles cached in memory by the operating system.

7.5.2.1 Random Tile Access Tests

For this test we will generate a single random list of tiles of levels 5 through 11. The list will contain 10,000 tile addresses. For each of the three file storage methods, we will iterate over the list of tiles and read each tile from disk. The code for the test is shown in Listing 7.2, and the results are shown in Table 7.3. In this test the single file per level method is fastest, but the database method is a close second. The single file per tile method is slowest.

	Single File per Tile	Single File per Level	Database Table per Level
Total Read Time (10,000 tiles)	379.455 seconds	112.357 seconds	146.926 seconds
Read Time per Tile	37.9 milliseconds	11.2 milliseconds	14.7 milliseconds

Table 7.3 Read times for random tile access.

7.5.2.2 Effect of Cached Tile Data

As stated earlier, modern operating systems cache disk file data in memory to speed up access. This test will demonstrate and measure the effect of such caching. In the previous test we read 10,000 random tiles from disk. In this test, we will read 1000 tiles 20 times. The first read will read from disk, and subsequent reads should pull from system memory.

Trial	Single File per Tile	Single File per Level	Database Table per Level
1	40.994	15.952	23.838
2	0.881	0.190	2.328
3	0.828	0.183	2.357
4	0.162	0.211	2.339
5	0.162	0.137	2.284
6	0.159	0.129	2.269
7	0.117	0.121	2.280
8	0.116	0.121	2.298
9	0.117	0.197	2.273
10	0.117	0.121	2.285
11	0.127	0.116	2.200
12	0.101	0.112	2.174
13	0.101	0.110	2.195
14	0.099	0.105	2.171
15	0.098	0.121	2.178
16	0.100	0.105	2.249
17	0.098	0.112	2.226
18	0.100	0.105	2.200
19	0.098	0.106	2.242
20	0.100	0.111	2.228

Table 7.4 Cached tile read times in seconds.

In Table 7.4, we can see that the first read of the 1000 tiles took by far the longest. Table 7.5 shows the results averaged with and without the first trial. We can see that the average times decreased significantly without the first trial.

	Single File per Tile	Single File per Level	Database Table per Level
Including first trial	2.2337	0.9232	3.3307
Excluding first trial	0.1937	0.1323	2.2514

Table 7.5 Average read times in seconds with and without first trial.

Table 7.6 compares the cached and non-cached tile read times. The single file per zoom level sees over an 8 to 1 improvement. The database table per zoom level sees over a 6 to 1 improvement. Finally, the single file per tile sees nearly a 20 to 1 improvement. In all cases, the single file per zoom level performs the best overall.

Consideration of memory cached tile files is important. In most cases the tiles from the top zoom levels will be the most commonly accessed, though they are the

	Single File per Tile	Single File per Level	Database Table per Level
No Caching	37.9	11.2	14.7
With Caching	1.9	1.3	2.2

Table 7.6 Cached versus non-cached tile read times in milliseconds.

lowest resolution. Tiled map clients will often start with a default view at the world or national level. Users will then zoom in to the specific areas they wish to view. Following this process will cause the top level tiles to be seen by almost all users. A very significant performance improvement can be realized by holding the most commonly accessed tiles in memory, either implicitly by the operating system or explicitly by the tile serving system.

7.6 Storage of Tile Metadata

So far we have not considered the need to store metadata about our source images and tiled images. Metadata includes all of the non-imagery data that might be needed. Important pieces of metadata that should be stored along side the tiled images include the date and version number of each tile and the source image(s) used to create each tile. Technical details like the resolution of source imagery used to make the tile or the original map level of the source data should also be included.

A system for maintaining tiled image sets should know which source images have been used to create which tiles so it can perform proper updates to those tiles when the source imagery changes. Large tiled image sets are often created from heterogeneous collections of imagery. Users of a tile-based system will want to know specifically what data was used to create the tiles.

This data is typically smaller than the image data. It can be stored in manners similar to storage of tiled images. In the case where we used a separate file for a tiled image, we could make a separate file for the metadata. We could also put the tile metadata in a database table or packed together in large files with tiled images. The specific means of storing of metadata is not as important as understanding and fulfilling the need to keep up with the data.

7.7 Storage of Tiles in Multi-Resolution Image Formats

The two key benefits of a tile-based system are that:

- Tiles are stored pre-rendered, exactly as needed for user consumption.
- Lower resolution views are pre-generated and quickly available.

The primary drawback of tile-based systems is the source imagery must undergo extensive reformatting. Multi-resolution image formats like JPEG2000 and MrSID are a possible alternative to this reformatting. As is, they meet one of the two key requirements for a tile based system. They use image transforms (typically wavelet based) to generate a multi-resolution encoding of an image. The multi-resolution views can be used to provide the lower resolution zoom level imagery for a tile-based system.

However, these formats do not meet the first requirement. To get a useable sub-image from a wavelet encoded image, several steps must be performed. Because the data is stored in multiple resolutions in multiple places in the file, several file seeks and reads are required. Use of these formats is a tradeoff. They eliminate the need for pre-processing and require less storage space, but they will always require more processing and I/O for tiled image retrieval.

7.8 Memory-Cached Tile Storage

In some cases tile retrieval performance must be as fast as possible. This can be a requirement to support real-time applications or to support many millions of users. In these cases, developers may want to create a method for caching entire tile sets in memory. Tile sets will still be archived to file but will be held in memory at run-time. Sizable tile sets will have to be spanned over several computers for this approach to work. Software systems like Memcached[2] are designed for exactly this type of problem. Memcached is used to cache large data sets in the memory of many separate computers.

7.9 Online Tile Storage

So far we have considered storing tiles in files on a computer's file system or memory. There are online file storage alternatives. Several services exist which allow Web accessible storage space to be rented. These services provide the storage space hosting with a high degree of reliability, often with multiple backups. One such service is Amazon's Simple Storage Service (S3)[3]. S3 is accessible through a web services interface that allows users to write and read data over HTTP. S3 uses a simple key-value storage system. Data objects (similar to BLOBs) are stored and accessible with a key. The key is used in the formation of an HTTP URL for access to the resource. For example, the following URL could be used to retrieve the binary resource.

`http://www.somestorageservice.com/mykey`

Since tiled images are discretely addressable and designed for use over HTTP, approaches like this are promising for tile-based systems. The primary disadvan-

---

[2] `http://memcached.org/`
[3] `http://aws.amazon.com/s3/`

tages of this type of storage will be cost and efficiency. However, for very large tile sets with many users, this type of system might be more cost-effective than the required hardware and bandwidth of a self-hosted solution.

Listing 7.1 Write test implementations.

```
1
2    public static void writeTileMultipleFiles(String outputFolder, int cols,
         int rows) {
3        File f = new File(outputFolder);
4        f.mkdirs();
5        byte[] data = new byte[byteSize];
6        for (int i = 0; i < cols; i++) {
7            String folderName = outputFolder + "/" + i;
8            File folder = new File(folderName);
9            folder.mkdirs();
10           for (int j = 0; j < rows; j++) {
11               File tileFile = new File(folderName + "/" + j + ".bin");
12               FileOutputStream fos;
13               try {
14                   fos = new FileOutputStream(tileFile);
15                   BufferedOutputStream bos = new BufferedOutputStream(fos);
16                   bos.write(data);
17                   bos.close();
18               } catch (Exception e) {
19                   e.printStackTrace();
20               }
21           }
22       }
23   }
24
25   public static void writeTilesSingleFile(String outputFolder, int cols, int
         rows, int level) {
26       File f = new File(outputFolder);
27       f.mkdirs();
28       byte[] data = new byte[byteSize];
29       IndexedTileOutputStream ptos = new IndexedTileOutputStream(f.
             getAbsolutePath(), "testing", new TileRange(0, cols − 1, 0, rows −
             1, level));
30       for (int i = 0; i < cols; i++) {
31           for (int j = 0; j < rows; j++) {
32               ptos.writeTile(i, j, data);
33           }
34       }
35       ptos.close();
36       String s = ptos.getBinFile();
37       IndexedTileInputStream iii = new IndexedTileInputStream(s);
38       iii.close();
39   }
40
41   public static void writeTileDatabase(String tableName, int cols, int rows,
         int level) {
42       try {
43           Connection c = DriverManager.getConnection("jdbc:postgresql://" + "
                 localhost/" + "tiledb", "user", "password");
44           Statement stmt;
45           stmt = c.createStatement();
46           byte[] data = new byte[byteSize];
47           try {
48               stmt.execute("DROP TABLE " + tableName);
49           } catch (Exception e) {
50           }
51           try {
52               stmt.execute("CREATE TABLE " + tableName + "(id bigserial
                     PRIMARY KEY ," + "row bigint , " + "col bigint," + "image
                     bytea )");
53           } catch (Exception e) {
54               e.printStackTrace();
55               return;
56           }
57           PreparedStatement ps = c.prepareStatement("INSERT INTO " +
                 tableName + "(row, col,image) VALUES (?,?,?)");
```

```
58          for (int i = 0; i < cols; i++) {
59              for (int j = 0; j < rows; j++) {
60                  ps.setLong(1, j);
61                  ps.setLong(2, i);
62                  ps.setBytes(3, data);
63                  ps.execute();
64              }
65          }
66          try {
67              stmt.execute("CREATE index " + tableName + "_index on " +
                    tableName + " (col,row)");
68          } catch (Exception e) {
69              e.printStackTrace();
70              return;
71          }
72      } catch (SQLException e1) {
73          e1.printStackTrace();
74      }
75  }
```

Listing 7.2 Random read tests.

```
1   private static void readTilesMultipleFiles(String dataLocation, ArrayList<
        String> lines, int trial, int numreads) {
2       int count = 0;
3       for (String s : lines) {
4           if (count == numreads) {
5               break;
6           }
7           count++;
8           String[] data = s.split(":");
9           String level = data[0];
10          String column = data[1];
11          String row = data[2];
12          String filename = dataLocation + "/" + trial + "_" + level + "/" +
                column + "/" + row + ".bin";
13          File f = new File(filename);
14          byte[] bytes = new byte[(int) f.length()];
15          try {
16              FileInputStream fis = new FileInputStream(f);
17              BufferedInputStream bis = new BufferedInputStream(fis);
18              DataInputStream dis = new DataInputStream(bis);
19              dis.readFully(bytes);
20              dis.close();
21              if (count % 1000 == 0) {
22                  System.out.println(count + ":" + bytes.length);
23              }
24          } catch (Exception e) {
25              e.printStackTrace();
26          }
27      }
28  }
29
30  private static void readTilesSingleFile(String dataLocation, ArrayList<
        String> lines, int trial, int numreads) {
31      IndexedTileInputStream[] streams = new IndexedTileInputStream[12];
32      int count = 0;
33      for (String s : lines) {
34          if (count == numreads) {
35              break;
36          }
37          count++;
38
39          String[] data = s.split(":");
40          int level = Integer.parseInt(data[0]);
41          long column = Long.parseLong(data[1]);
```

```
42      long row = Long.parseLong(data[2]);
43      if (streams[level] == null) {
44          long maxCol = TileStandards.zoomColumns[level] - 1;
45          long maxRow = TileStandards.zoomRows[level] - 1;
46          streams[level] = new IndexedTileInputStream(dataLocation + "/"
                + trial + "_" + level, "testing", level);

47
48      }
49      IndexedTileInputStream itis = streams[level];
50      byte[] bytes = itis.getTile(column, row);
51
52  }
53  for (int i = 0; i < streams.length; i++) {
54      if (streams[i] != null) {
55          streams[i].close();
56      }
57  }
58
59  }
60
61  private static void readTilesDatabase(ArrayList<String> lines, int numreads
        ) {
62      try {
63          Connection c = DriverManager.getConnection("jdbc:postgresql://" + "
                localhost/" + "tiledb", "username", "password");
64          Statement stmt;
65          stmt = c.createStatement();
66          int count = 0;
67          for (String s : lines) {
68              if (count == numreads) {
69                  break;
70              }
71              count++;
72              String[] data = s.split(":");
73              int level = Integer.parseInt(data[0]);
74              long column = Long.parseLong(data[1]);
75              long row = Long.parseLong(data[2]);
76              String tableName = "tiles_" + level;
77              ResultSet rs = stmt.executeQuery("SELECT image from " +
                    tableName + " where col=" + column + " and row=" + row);
78              rs.next();
79              byte[] bytes = rs.getBytes(1);
80          }
81          stmt.close();
82      } catch (SQLException e1) {
83          e1.printStackTrace();
84      }
85
86  }
```

Chapter 8
Practical Tile Storage

The previous chapter gave overviews for several different methods for storing tiled images. In this chapter we will present two fully-implemented techniques for storing tiled images together in large files. This type of method proved to be the best performing for writing, random reading, cached reading, and bulk copying. Furthermore, it is rather simple to implement. The first implementation shown is a fully functional method for writing and reading tile files and takes only about 200 lines of Java code for the reading, writing, and indexing methods.

Additionally, we will present the techniques with accompanying methods for creating tile indexes. These storage methods are designed to handle large and small sets of tiled images and are portable and updateable.

8.1 Introduction to Tile Indexes

Our goal is to store many hundreds or thousands of tiled images in a single file. This could be done by simply writing each image sequentially to a file. However, there would be no way to retrieve the images individually. There would be no way to know which tile address corresponded to which image. We could store the tile address before each image in our file, as shown in Figure 8.1. The problem with this method should be obvious. In order to access a specific tiled image, we have to scan the whole file. For tiled images sets of any significant size this method would be prohibitively inefficient. Instead we need to create a separate index into the file that will allow us to quickly look up the location of a specific tile in the file.

There are two principal ways we can construct the index:

- Sequential list of tile address to file position pairs
- Direct lookup table of file positions.

The simplest method is just to store the tile address, the position in the file and the size of the tiled image in a sequential list. This method is shown in Figure 8.2. The sequential list must still be searched for each tile query. However, the index

J.T. Sample and E. Ioup, *Tile-Based Geospatial Information Systems:*
Principles and Practices, DOI 10.1007/978-1-4419-7631-4_8,
© Springer Science+Business Media, LLC 2010

Record 1	Tile Address (Row, Column, Level)	Tile Image (N bytes)
Record 2	Tile Address (Row, Column, Level)	Tile Image (N bytes)
Record 3	Tile Address (Row, Column, Level)	Tile Image (N bytes)
Record 4	Tile Address (Row, Column, Level)	Tile Image (N bytes)

Fig. 8.1 Tile file with embedded addresses.

Index File

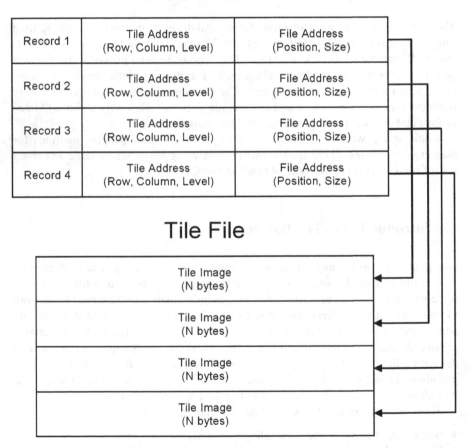

Record 1	Tile Address (Row, Column, Level)	File Address (Position, Size)
Record 2	Tile Address (Row, Column, Level)	File Address (Position, Size)
Record 3	Tile Address (Row, Column, Level)	File Address (Position, Size)
Record 4	Tile Address (Row, Column, Level)	File Address (Position, Size)

Tile File

Tile Image (N bytes)
Tile Image (N bytes)
Tile Image (N bytes)
Tile Image (N bytes)

Fig. 8.2 Tile index stored as sequential list of addresses.

information is several orders of magnitude smaller than the actual tiles, thus scanning a separate index file for each query, while still inefficient, is must faster than scanning the whole file. An optimized variant of this method is to sort the data in the tile index. In this fashion a linear search could be used to speed up searching the index list.

The second method is to create a direct lookup table of file positions. For example, as shown in Figure 8.3, tile zoom level 3 has 8 columns and 4 rows, so the lookup table only requires 32 records. The table could be stored in a file in row-major or column-major format. The position of any given record can be directly computed, and only a single seek and read is required to retrieve the file position for a specific tile. We can store a null value in the lookup table to indicate that a tile does not exist for that specific address. If the table is stored in row-major order,

Lookup Table: Zoom Level 3

	Column 0	Column 1	Column 2	Column 3	Column 4	Column 5	Column 6	Column 7
Row 0	File Address (Position, Size)	File Address (Position, Size)	File Address (Position, Size)	File Address (Position, Size)	File Address (Position, Size)	File Address (Position, Size)	File Address (Position, Size)	File Address (Position, Size)
Row 1	File Address (Position, Size)	File Address (Position, Size)	File Address (Position, Size)	File Address (Position, Size)	File Address (Position, Size)	File Address (Position, Size)	File Address (Position, Size)	File Address (Position, Size)
Row 2	File Address (Position, Size)	File Address (Position, Size)	File Address (Position, Size)	File Address (Position, Size)	File Address (Position, Size)	File Address (Position, Size)	File Address (Position, Size)	File Address (Position, Size)
Row 3	File Address (Position, Size)	File Address (Position, Size)	File Address (Position, Size)	File Address (Position, Size)	File Address (Position, Size)	File Address (Position, Size)	File Address (Position, Size)	File Address (Position, Size)

Fig. 8.3 Lookup table for zoom level 3.

Equation 8.1 is used to compute the position in the array of addresses.

$$p = j * C + i \tag{8.1}$$

where

$$i = \text{column index}$$
$$j = \text{row index}$$
$$C = \text{number of columns}$$
$$p = \text{position of tile record}$$

The disadvantage of this approach is that the size of our lookup table file grows by 4 times for each successive level. If the file address is stored as an 8 byte integer and the size is stored as a 4 byte integer, we need 12 bytes for each record. Zoom level 17 contains 131,072 columns and 65,536 rows for a total of 8,589,934,592 tiles. This would require over 100 gigabytes just for the index file. If we had a tile set with a complete (or nearly complete) coverage of the earth's surface at that resolution, this approach would be appropriate.

However, this is unlikely. Most of the earth's surface is covered with water (liquid and ice) that is rarely imaged at high resolution. Few tile sets will cover even a fraction of the earth's surface. In these cases, we should develop an indexing method that provides direct lookup of tile locations, but also allows us to have lookup tables that cover only a subset of the entire level. This can be easily accomplished by providing for offsets attached to the index table. Rather than having all index tables start at (0,0) and covering the full range of tile addresses, we can provide external start and end addresses for index tables.

The next two sections will each present an algorithm for storing large amounts of tiled images. Each algorithm comes with its own unique method for indexing tiles. Those methods are modified versions of the direct lookup algorithm.

8.2 Storage by Zoom Level

Our first technique for storing tiles is to store all the tiles for a specific zoom level in a single file. This is the same technique that was tested and benchmarked in the previous chapter. This technique uses three files for each zoom level, one file for the tiled images and two files for the index.

The file containing the tiled images is simply a sequential list of tiled images. It first stores a magic number to serve as a sentinel value. Then it stores the tile's address and size. Finally it stores the tiled image data. The sentinel values and tile addresses are stored to make the tile images recoverable in the case that the tile file or index files become corrupted. Figure 8.4 shows the record structure for the tiled image file. Since the tiles do not have to be stored in any particular order, tiles can be written over a period of time. New tiles can be added to the file by simply writing them at the end of the file.

The index storage is slightly more complicated. Recall from the previous section that our lookup table based method can require a very large lookup table for the higher resolution zoom levels. To reduce the required size we have designed a two-step lookup table. We use the same approach to writing the lookup table from Figure 8.3, except that we only store rows in the index file that actually have tiles in them. So if our tile set only has 100 rows, then our tile index will only have 100 rows worth of tile addresses.

To accomplish this we have to create an additional index file, a row index file. This file contains a single value for each row in our set. If the row has any tiles, we store the location of that row's index records from the tile index file. If the row does not have any tiles, we store a null value in the file.

Magic Number	Tile Address (Row, Column)	Tile Size	Tile Image (N bytes)
Magic Number	Tile Address (Row, Column)	Tile Size	Tile Image (N bytes)
Magic Number	Tile Address (Row, Column)	Tile Size	Tile Image (N bytes)
Magic Number	Tile Address (Row, Column)	Tile Size	Tile Image (N bytes)

Fig. 8.4 Structure of tiled image file.

An example of this method is shown in Figure 8.5. We have used the same table from Figure 8.3, but we have assumed that rows 0 and 2 contain zero tiles. In this case, neither of those rows is stored in the index table, and the subsequent table is only half the size. Thus, the advantage of this technique is reduced space requirements for the index file. The disadvantage is that we have to do two seeks and reads to get the tile address. However, as shown in the previous chapter's benchmarks, the performance is still very good.

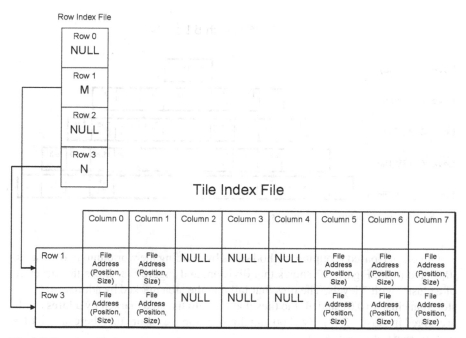

Fig. 8.5 Two-step tile index method.

To get the address for a specific tile, seek to the position of the row pointer in the row index file and read the value. If the value is non-null, use that value to position the tile index file. Then seek additional positions for the column index and read the tile address. Listings 8.1 and 8.2 present example code for writing and reading indexed tiles.

8.3 Introduction to Tile Clusters

The previous method works well and could be modified such that all levels can be contained in a single file. This would require addition of a third index file, a level index file similar to the row index file. Each tile address lookup would require 3 seeks and reads.

However, this method would not address two of the problems discussed in the tile creation chapter. Recall both the performance improvements made possible by caching tiles in memory (Section 6.1) and the requirement to have logically defined sub-groupings of tiles for distributed tile creation (Section 6.3.2). To address both of these requirements we propose a method for grouping tiled images in clusters.

Tiled image layers follow a pyramid type structure, see Figure 8.6. Each level has 4 times the number of tiles as its predecessor. Also, each lower resolution level is based on the image data from the next higher resolution level.

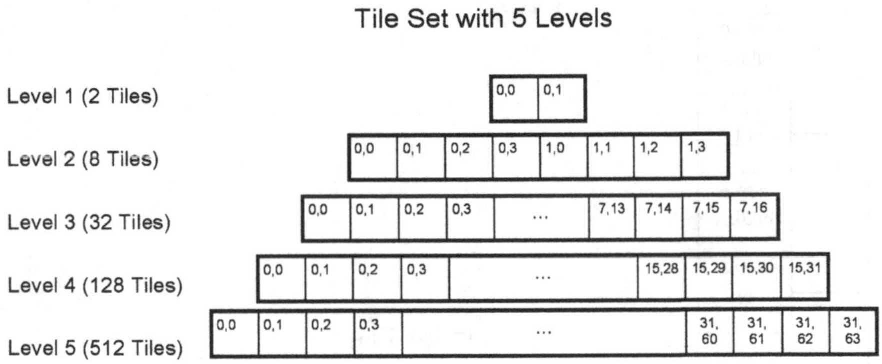

Tile Set with 5 Levels

Fig. 8.6 Pyramid structure of tile images.

Our cluster-based grouping method starts by dividing the world into two clusters, (0,0) and (0,1). Figure 8.7 shows that division, and Figure 8.8 shows the structure of a cluster with 5 levels. The tiles that fall into the area marked by address (0,0) are stored in cluster (0,0), and all the tiles that fall into the area marked by address (0,1) are stored in cluster (0,1). By choosing this division we ensure that there are no tiles that overlap both clusters.

The number of tiles for a tile set with l levels is computed with Equation 8.2:

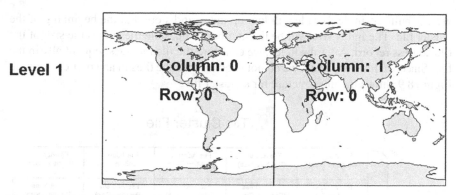

Fig. 8.7 World divided into two clusters.

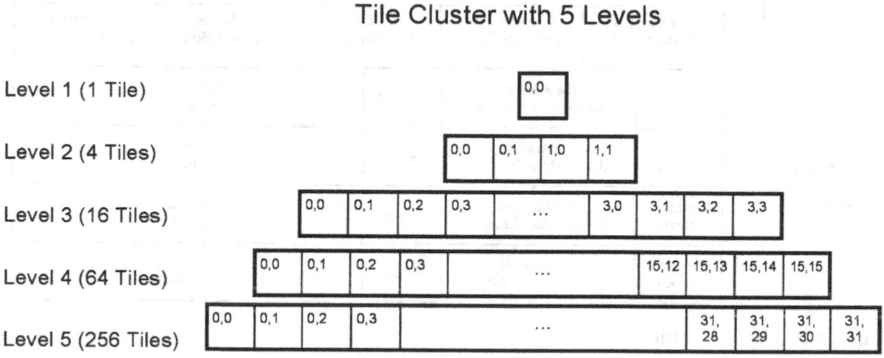

Tile Cluster with 5 Levels

Level 1 (1 Tile)	0,0

Level 2 (4 Tiles)	0,0	0,1	1,0	1,1

Level 3 (16 Tiles)   0,0 | 0,1 | 0,2 | 0,3 | ... | 3,0 | 3,1 | 3,2 | 3,3

Level 4 (64 Tiles)   0,0 | 0,1 | 0,2 | 0,3 | ... | 15,12 | 15,13 | 15,14 | 15,15

Level 5 (256 Tiles)   0,0 | 0,1 | 0,2 | 0,3 | ... | 31,28 | 31,29 | 31,30 | 31,31

Fig. 8.8 Structure of a cluster with 5 levels.

$$N = \sum_{i=1}^{L} 2^i 2^{i-1} \tag{8.2}$$

The number of tiles for a cluster with l levels is the value from Equation 8.2 divided by two, or as shown in Equation 8.3:

$$N' = \sum_{i=1}^{L} 2^{2i-2} \tag{8.3}$$

8.4 Tile Cluster Files

To store tiles in cluster files, we must first set the number of levels to be stored. For a tile set with a base level of 7, we will need two cluster files, each with 7 levels of tiles and 5,461 tiles. Because the possible number of tiles is fixed for each cluster, we can build a single fixed length lookup index and store it at the beginning of the cluster file. The index size will be the number of possible tiles times the size of the tile address record. After the index, we can store the tiled images sequentially in the file. Since we have an index, we do not need to store the tiles in any particular order. Figure 8.9 shows the file structure for a cluster file.

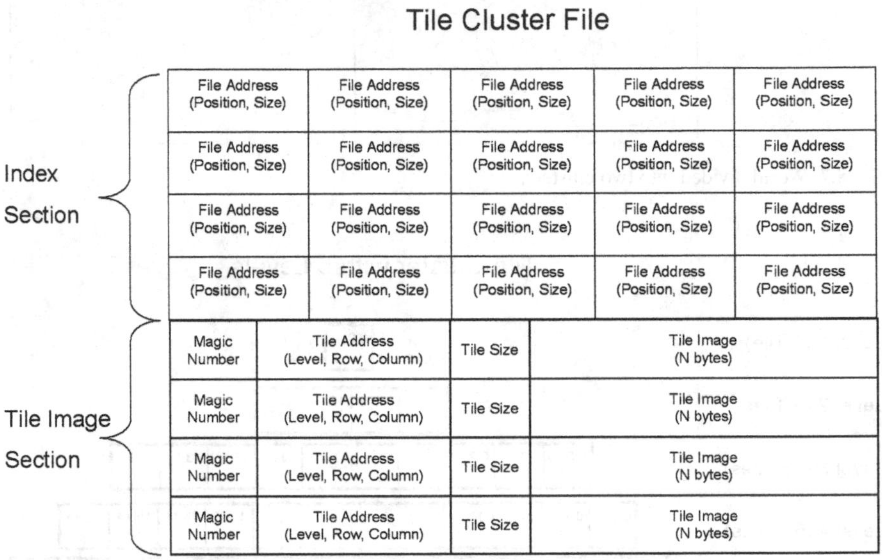

Tile Cluster File

File Address (Position, Size)	File Address (Position, Size)	File Address (Position, Size)	File Address (Position, Size)	File Address (Position, Size)
File Address (Position, Size)	File Address (Position, Size)	File Address (Position, Size)	File Address (Position, Size)	File Address (Position, Size)
File Address (Position, Size)	File Address (Position, Size)	File Address (Position, Size)	File Address (Position, Size)	File Address (Position, Size)
File Address (Position, Size)	File Address (Position, Size)	File Address (Position, Size)	File Address (Position, Size)	File Address (Position, Size)

Index Section (brace covering the rows above)

Magic Number	Tile Address (Level, Row, Column)	Tile Size	Tile Image (N bytes)
Magic Number	Tile Address (Level, Row, Column)	Tile Size	Tile Image (N bytes)
Magic Number	Tile Address (Level, Row, Column)	Tile Size	Tile Image (N bytes)
Magic Number	Tile Address (Level, Row, Column)	Tile Size	Tile Image (N bytes)

Tile Image Section (brace covering the rows above)

Fig. 8.9 Structure of a tile cluster file.

8.5 Multiple Levels of Clusters

When applying this method to tile sets with several more than 7 levels, we will experience the same problem discussed in Section 8.1. Our index will be too large. Imagine a tile set with only 100 tiles at level 15. Scaled versions of those 100 tiles will give us about 50 additional tiles with levels 14 to 1. That is a total of 150 tiles. If each tile is 50,000 bytes then the size of the tiles in total is 7.5 megabytes. However, a cluster file with 15 levels can have up to 357,913,941 tiles. If each index record takes up 12 bytes, the size of the index table would be 4,294 megabytes, or almost 600 times the size of the actual image data. This is a highly impractical consequence.

To alleviate this problem we allow multiple levels of clusters, with each level covering a continuous sub-range of levels. For example, if we have a tile set with 15 levels, we will have two levels of clusters, one level with contain tile levels 1-7, and the other level of clusters will contain levels 8-15. The first level contains 7 levels, and the second level contains 8 levels. The indexes for multi-level cluster groups will never grow unmanageably large.

Continuing the example of a multi-level set of clusters, the first set, those with levels 1-7, can only have up to two clusters. While the second set, representing levels 8-15 can contain as many clusters as there are tiles in level 8. This number is 32,768. However, in practice we will only create clusters files when there are tiles that belong in the cluster. Few tile sets will have complete coverage of the whole world at a high resolution, and thus the full 32,768 would never actually be needed. The actual required number would fluctuate based on the size of the tile set.

Multi-Level Cluster Structure

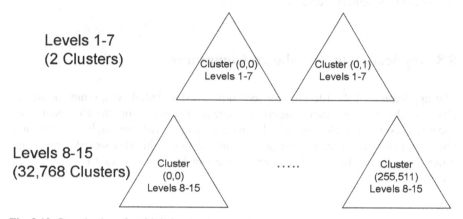

Fig. 8.10 Organization of multiple levels of tile clusters.

8.6 Practical Implementation of Tile Clusters

Listing 8.3 implements a cluster-based tile storage method. Since the internal file structure of our cluster files is relatively simple the implementation is relatively straightforward. The most difficult component of the practical implementation of our cluster-based storage system is the algorithm to determine in which cluster a given tile should be placed. That algorithm can be seen in the methods "getIndex-Position" and "getClusterFileTileAddress".

8.7 Application to Memory Cached Tiles

In Chapter 5, we saw potential performance improvements from holding tiles in memory while they were being created. The cluster-based storage technique works very well with this concept. To implement this with the cluster technique, we first ensure that our clusters are divided small enough to be held uncompressed completely in memory. If that is the case, then we alter our tile creation method to create tiles for one cluster at a time. We modify our clustered tile storage algorithm by simply adding a cache that holds all the tiles in memory as they are written. It writes them to file at the end of the tile creation process. This offers an additional performance improvement. If we write all the tiles at one time, we can write them sequentially and avoid using random file seeks, reads and writes. Random file accesses are generally slower than sequential accesses.

8.8 Application to Distributed Computing

The application of the tile clusters technique to distributed computing should be obvious. Multiple computer systems can be tasked with creating the tiles locally for specific cluster files. The individual cluster files can exist separately and function with minimal interaction, so they are a natural fit for highly distributed computing. After a cluster is completed, the single file can be copied back to a centralized location.

8.9 Performance Optimizations of Tile Cluster Method

There are several other potential performance optimizations available with the clustered storage technique. First, in our example code we opened and closed the various cluster files for each read and write. This is generally slower than maintaining constantly open files and reading and writing from them. Therefore, we might alter our algorithm to keep all the cluster files open throughout the process. However, many systems enforce a limit on the number of open files at any one time. Therefore, to get some performance benefit we can maintain a cache of recently opened files. The cache needs to be of sufficient size to ensure that open files are reused, but it must not be larger than the allowed maximum number of open files.

Since new tiles are written at the end of the file, this technique supports adding tiles over a period of time. When an existing tile is overwritten, the index is updated to point to the new tile. Old tiles remain in the file and take up space. Some developers may want to implement a system to try to re-use that space, either by trying to fit other tile images in the old space or by periodically rebuilding the entire cluster.

Finally, when tiles are served to users from clusters the performance can be quite good. Users typically view tiles for a specific area, and with our system those tiles would be clustered in the same file. However, there is one case where the performance can be rather poor. Recall our example in which we had tiles in levels 1-15, and separated the clusters into groups of 1-7 and 8-15 levels. If a user is viewing level 8, and requesting several tiles from level 8, the system will have to access a different file for each tile. The benchmarks in the previous chapter showed that using a separate file per tile can be somewhat slow. A workaround to this problem is to build in some overlap in our cluster structure. Instead of a 1-7 and 8-15 break, we will use a 1-8 and 8-15 break. The tiles from level 8 are stored in two places. This does introduce some inefficiency; level 8 can have up to 32,768 tiles. But the read performance improvements may be worth the cost.

Listing 8.1 Output of indexed tiles by zoom level.

```java
public class IndexedTileOutputStream {

    static final long magicNumber = 0x772211ee;

    String imagefilename;
    String rowindexname;
    String tileindexname;

    RandomAccessFile imagefile;
    RandomAccessFile rowindex;
    RandomAccessFile tileindex;

    long numrows, numcolumns;
    int rowRecordSize = 8;
    int tileRecordSize = 8 + 4;

    public IndexedTileOutputStream(String folder, String setname, int level
            ) {
        imagefilename = folder + "/" + setname + "-" + level + ".tiles";
        rowindexname = folder + "/" + setname + "-" + level + ".rowindex";
        tileindexname = folder + "/" + setname + "-" + level + ".tileindex"
            ;
        numrows = TileStandards.zoomRows[level];
        numcolumns = TileStandards.zoomColumns[level];

        try {
            imagefile = new RandomAccessFile(imagefilename, "rw");

            // if the row file is empty, fill it with -1 to indicate empty
                values
            rowindex = new RandomAccessFile(rowindexname, "rw");
            if (rowindex.length() == 0) {
                rowindex.seek(0);
                for (int i = 0; i < numrows; i++) {
                    rowindex.writeLong(-1L);
                }
            }
            tileindex = new RandomAccessFile(tileindexname, "rw");
        } catch (Exception e) {
            e.printStackTrace(),
        }
    }

    public void writeTile(long col, long row, byte[] data) {

        try {
            // position tile file to write at end of file
            long writepos = imagefile.length();
            imagefile.seek(writepos);

            // write tile address and imagedata to file
            // write two magic numbers so that tile records can be recovered
                in case of corrupted file
            imagefile.writeLong(magicNumber);
            imagefile.writeLong(magicNumber);
            imagefile.writeLong(col);
            imagefile.writeLong(row);
            imagefile.writeInt(data.length);
            imagefile.write(data);

            // update index
            updateIndex(col, row, writepos, data.length);

        } catch (IOException e) {
            e.printStackTrace();
```

```
63            }
64
65        }
66
67        private void updateIndex(long col, long row, long writepos, int length)
          {
68            try {
69                //check if row is in the row index
70                long rowposition = rowRecordSize * row;
71                rowindex.seek(rowposition);
72                long rowpointer = rowindex.readLong();
73                if (rowpointer == -1L) {
74                    //this means the row data is new and not already in the
                        index
75                    rowpointer = tileindex.length();
76                    tileindex.seek(rowpointer);
77                    //write an array of empty values
78                    for (int i = 0; i < numcolumns; i++) {
79                        tileindex.writeLong(-1L);
80                        tileindex.writeInt(-1);
81                    }
82                    //write the position back to the original row index
83                    rowindex.seek(rowposition);
84                    rowindex.writeLong(rowpointer);
85                }
86                //compute offset into row for specific col
87                long offset = rowpointer + col * tileRecordSize;
88                //position tile index for writing the file address of the tile
                    image
89                tileindex.seek(offset);
90                tileindex.writeLong(writepos);
91                tileindex.writeInt(length);
92            } catch (IOException e) {
93                e.printStackTrace();
94            }
95        }
96
97        public void close() {
98            try {
99                imagefile.close();
100               rowindex.close();
101               tileindex.close();
102           } catch (Exception e) {
103           }
104       }
105
106   }
```

Listing 8.2 Reading indexed tiles.

```
1
2  public class IndexedTileInputStream {
3
4          String imagefilename;
5          String rowindexname;
6          String tileindexname;
7
8          RandomAccessFile imagefile;
9          RandomAccessFile rowindex;
10         RandomAccessFile tileindex;
11
12         long numrows, numcolumns;
13         int rowRecordSize = 8;
14         int tileRecordSize = 8 + 4;
15
```

```
16        public IndexedTileInputStream(String folder, String setname, int level)
              {
17            imagefilename = folder + "/" + setname + "-" + level + ".tiles";
18            rowindexname = folder + "/" + setname + "-" + level + ".rowindex";
19            tileindexname = folder + "/" + setname + "-" + level + ".tileindex"
              ;
20            numrows = TileStandards.zoomRows[level];
21            numcolumns = TileStandards.zoomColumns[level];
22
23            try {
24                imagefile = new RandomAccessFile(imagefilename, "rw");
25                rowindex = new RandomAccessFile(rowindexname, "rw");
26                tileindex = new RandomAccessFile(tileindexname, "rw");
27            } catch (Exception e) {
28                e.printStackTrace();
29            }
30        }
31
32        public byte[] getTile(long col, long row) {
33            try {
34                //check if row is in the row index
35                long rowposition = rowRecordSize * row;
36                rowindex.seek(rowposition);
37                long rowpointer = rowindex.readLong();
38                if (rowpointer == -1L) {
39                    //this means the row data is not in the index, and so the
                      tile doesn't exist
40                    return null;
41                }
42                //compute offset into row for specific col
43                long offset = rowpointer + col * tileRecordSize;
44                //position tile index for reading the position and size of the
                  tile image
45                tileindex.seek(offset);
46                long tileposition = tileindex.readLong();
47                int size = tileindex.readInt();
48                if (tileposition == -1L) {
49                    //this means that the tile isn't there
50                    return null;
51                }
52                //adjust the tile position to skip the magic numbers and
                  address information
53                long adjustedTilePosition = tileposition + 8 + 8 + 8 + 8 + 4;
54                byte[] data = new byte[size];
55                //position the image file and read the image data
56                imagefile.seek(adjustedTilePosition);
57                imagefile.readFully(data);
58                return data;
59            } catch (IOException e) {
60                e.printStackTrace();
61            }
62            return null;
63        }
64
65    }
```

Listing 8.3 Tile clusters implementation.

```
1
2    public class ClusteredTileStream {
3
4        static final long magicNumber = 0x772211ee;
5        private String location;
6        private String setname;
7        private int numlevels;
8        private int breakpoint;
```

```
 9
10      public ClusteredTileStream(String location, String setname, int
            numlevels, int breakpoint) {
11          this.location = location;
12          this.setname = setname;
13          this.numlevels = numlevels;
14          this.breakpoint = breakpoint;
15      }
16
17      public void writeTile(long row, long column, int level, byte[]
            imagedata) {
18
19          //first determine the cluster that will hold the data
20          ClusterAddress ca = getClusterForTileAddress(row, column, level);
21          String clusterFile = getClusterFileForAddress(ca);
22          if (clusterFile == null) {
23              return;
24          }
25          File f = new File(clusterFile);
26
27          //if the file doesn't exist, set up an empty cluster file
28          if (!f.exists()) {
29              createNewClusterFile(f, ca.endlevel - ca.startlevel + 1);
30          }
31          try {
32              RandomAccessFile raf = new RandomAccessFile(f, "rw");
33
34              //write the tile and info at the end of the tile file
35              long tilePosition = raf.length();
36              raf.seek(tilePosition);
37              raf.writeLong(magicNumber);
38              raf.writeLong(magicNumber);
39              raf.writeLong(column);
40              raf.writeLong(row);
41              raf.writeInt(imagedata.length);
42              raf.write(imagedata);
43
44              //determine the position in the index of the tile address
45              long indexPosition = getIndexPosition(row, column, level);
46              raf.seek(indexPosition);
47
48              //write the tile position and size in the index
49              raf.writeLong(tilePosition);
50              raf.writeInt(imagedata.length);
51              raf.close();
52          } catch (Exception e) {
53              e.printStackTrace();
54          }
55      }
56
57      public byte[] readTile(long row, long column, int level) {
58          //first determine the cluster that will hold the data
59          ClusterAddress ca = getClusterForTileAddress(row, column, level);
60          String clusterFile = getClusterFileForAddress(ca);
61          if (clusterFile == null) {
62              return null;
63          }
64          File f = new File(clusterFile);
65
66          try {
67              RandomAccessFile raf = new RandomAccessFile(f, "r");
68
69              //determine the position in the index of the tile address
70              long indexPosition = getIndexPosition(row, column, level);
71              raf.seek(indexPosition);
72              long tilePosition = raf.readLong();
73              int tileSize = raf.readInt();
```

```
74            if (tilePosition == -1L) {
75                // tile is not in the cluster
76                raf.close();
77                return null;
78            }
79            byte[] imageData = new byte[tileSize];
80            // offset tile position for header information
81            long tilePositionOffset = tilePosition + 8 + 8 + 8 + 8 + 4;
82            raf.seek(tilePositionOffset);
83            raf.readFully(imageData);
84            raf.close();
85            return imageData;
86        } catch (Exception e) {
87            e.printStackTrace();
88        }
89        return null;
90    }
91
92    private long getIndexPosition(long row, long column, int level) {
93        ClusterAddress ca = this.getClusterForTileAddress(row, column,
                level);
94        // compute the local address, that's the relative address of the
                tile in the cluster
95        int locallevel = level - ca.startlevel;
96        long localRow = (long) (row - (Math.pow(2, locallevel) * ca.row));
97        long localColumn = (long) (column - (Math.pow(2, locallevel) * ca.
                column));
98        int numColumnsAtLocallevel = (int) Math.pow(2, locallevel);
99        long indexPosition = this.getCumulativeNumTiles(locallevel - 1) +
                localRow * numColumnsAtLocallevel + localColumn;
100       // multiply index position times byte size of a tile address
101       indexPosition = indexPosition * (8 + 4);
102       return indexPosition;
103    }
104
105    public ClusterAddress getClusterForTileAddress(long row, long column,
            int level) {
106        if (level > this.numlevels) {
107            // error, level is outside of ok range
108            return null;
109        }
110        int targetLevel = 0;
111        int endLevel = 0;
112        if (level < breakpoint) {
113            // tile goes in one of top two clusters
114            targetLevel = 1;
115            endLevel = breakpoint - 1;
116        } else {
117            // tile goes in bottom cluster
118            targetLevel = this.breakpoint;
119            endLevel = this.numlevels;
120        }
121        // compute the difference between the target cluster level and the
                tile level
122        int powerDiff = level - targetLevel;
123        // level factor is the number of tiles at level "level" for a
                cluster that starts at "target level"
124        double levelFactor = Math.pow(2, powerDiff);
125        // divide the row and column by the level factor to get the row and
                column address of the cluster we are using
126        long clusterRow = (int) Math.floor(row / levelFactor);
127        long clusterColumn = (int) Math.floor(column / levelFactor);
128        ClusterAddress ca = new ClusterAddress(clusterRow, clusterColumn,
                targetLevel, endLevel);
129        return ca;
130    }
131
```

```
132    String getClusterFileForAddress (ClusterAddress ca) {
133        String filename = this.location + "/" + this.setname + "−" + ca.
               startlevel + "−" + ca.row + "−" + ca.column + ".cluster";
134        return filename;
135    }
136
137    //this methods create an empty file and fills the index with null
           values
138    void createNewClusterFile (File f, int numlevels) {
139        RandomAccessFile raf;
140        try {
141            raf = new RandomAccessFile (f, "rw");
142            raf.seek (0);
143            long tiles = this.getCumulativeNumTiles (numlevels);
144            for (long i = 0; i < tiles; i++) {
145                raf.writeLong(−1L); //NULL position of tile
146                raf.writeLong(−1L); //NULL size of tile
147            }
148            raf.close ();
149        } catch (Exception e) {
150            e.printStackTrace ();
151        }
152    }
153
154    public int getCumulativeNumTiles (int finallevel) {
155        int count = 0;
156        for (int i = 1; i <= finallevel; i++) {
157            count += (int) (Math.pow(2, 2 * i − 2));
158        }
159        return count;
160    }
161
162    }
163
164    public class ClusterAddress {
165
166        long row;
167        long column;
168        int startlevel;
169        int endlevel;
170
171        public ClusterAddress(long row, long column, int startlevel, int
               endlevel) {
172            this.row = row;
173            this.column = column;
174            this.startlevel = startlevel;
175            this.endlevel = endlevel;
176        }
177
178    }
```

Chapter 9
Tile Serving

The previous four chapters explained techniques for creating and storing tiled images. In this chapter we will examine methods for sharing those tiled images with other users over a network.

9.1 Basics of HTTP

The Hypertext Transfer Protocol (HTTP) is one of the core standards of the Internet. It was originally designed for sharing interlinked documents but is now used for many other types of applications. Since HTTP is the basic application protocol of the Internet, there are considerable existing tools in place to support it. HTTP clients and/or servers are built into many applications and programming environments. Commercial and residential firewalls and proxies are designed to handle HTTP traffic. HTTP has a sophisticated security model with extensive infrastructure. What is most important to our designed of a tiled image server is that HTTP can be used to efficiently and securely share tiled images to both Web browsers and other applications.

The HTTP standard defines eight operations, but only two will be relevant to our work: GET and POST. The GET operation's intended use is to retrieve web content without causing side effects on the server. Consider a simple service to retrieve the current time from a time server. This would be best modeled with a GET operation, because there is no need to change any information on the time server. The POST operation is used to "post" data back to the server. For example, a web service that processed book orders would be modeled with the POST operation. This operation does cause side effects on the server because repeated executions of the post operation will result in multiple book orders. Since we are concerned with serving tiled images to users, not collecting data from users, we should use the "GET" operation to create our tiled image service. HTTP GET encodes query parameters by concatenating them on the end of the resource's Uniform Resource Locator (URL). For example, a URL to a time service might be:

J.T. Sample and E. Ioup, *Tile-Based Geospatial Information Systems:* 151
Principles and Practices, DOI 10.1007/978-1-4419-7631-4_9,
© Springer Science+Business Media, LLC 2010

```
http://www.sometimeservice.com/getthetime
```

Entering this URL into a web browser would return the current time encoded as an HTML document. Suppose that you want to retrieve the time for a different time zone. You could add a parameter called "zone" to send the server a time zone you want. The resulting URL would look like this:

```
http://www.sometimeservice.com/getthetime?zone=UTC-6
```

An additional parameter could be added to specify the encoding for the response like this:

```
http://www.sometimeservice.com/getthetime?zone=
UTC-6&encoding=XML
```

9.2 Basic Tile Serving

In general, serving tiles requires a multi-step process. In the first step, users query a tile server of a list of available layers. They may also query the server for the availability of specific tiles from each layer and the image format of the specific tiles. In the final step, users request the actual tiled images.

In practice the interaction may be much simpler. Clients will usually not ask for the format of a tile. We have mandated that our tiled images be stored in browser compatible formats, either JPEG or PNG, so web browsers will be able to read them without querying first to see what format they use. Also, rather than use two queries to check if a tile is there and then to request it, clients will just request the tile, and accept the response that comes back. Many servers will provide a default tile to stand in for missing ones. This can be an empty, completely transparent tile, or a tile with a message that says the tile is not available.

The next two sections will present two slightly different schema for serving tiled images. They differ in the manner in which tiles are queried. The reader should not infer that the two different schema are incompatible or must be implemented on different systems. Each scheme is merely an interface to the same backing data store. Both schemes can be implemented on the same server. In practice, many different interfaces are implemented on tile stores to support as many clients as possible. Modern HTTP servers and Web service frameworks allow multiple interfaces to be implemented with very minimal overhead. In the final chapter, we will look at other methods for serving tiles to accommodate multiple client software packages and standardized protocols.

Each of the following algorithms will make use of the data store concept. The data store provides tiled images organized into layers. Individual tiles are queried by

Listing 9.1 Java DataStore abstract class.

```java
public abstract class DataStore {

    // returns a list of layers availible from this DataStore
    public abstract String[] getLayersAvailible();

    // returns a list of TileRanges for the zoom levels availible from the named
    //     layer
    // null values in the array indicate a missing zoom level
    public abstract TileRange[] getTileRanges(String layerName);

    public abstract boolean tileExists(String layerName, int level, long row,
        long column);

    public abstract String getTileFormat(String layerName, int level, long row,
        long column);

    public abstract byte[] getTileImage(String layerName, int level, long row,
        long column);

}

class TileRange {

    int level;
    long mincol;
    long maxcol;
    long minrow;
    long maxrow;

}
```

layer, zoom level, row and column. The data store also provides information about the layers, such as what zoom levels and tile ranges are available for each layer, and whether or not a tile exists for a given address. Java and Python abstractions for the data store concept are shown in Listings 9.1 and 9.2

9.3 Tile Serving Scheme with Encoded Parameters

This section presents a tile scheme in which all parameters needed to query the tile server are encoded as HTTP GET parameters concatenated to the tile server's URL. A URL similar to the following will serve as the base URL for all tile requests:

`http://www.sometileserver.com/tiles`

If left without any parameters it will return a simple text-encoded list of available layers, the zoom levels and ranges of tiles available for each layer. The layer list response will appear as in Tables 9.1 and 9.2. This list can easily be parsed by client applications. Tile request parameters are given in Table 9.3.

Listing 9.2 Python DataStore abstract class

```
1   class DataStore:
2       def getLayersAvailable(self):
3           return []
4
5       def getTileRanges(self):
6           return []
7
8       def tileExists(self):
9           return False
10
11      def getTileFormat(self):
12          return ''
13
14      def getTileImage(self):
15          return ''
16
17  class TileRange:
18      def __init__(self):
19          self.level = 0
20          self.mincol = 0
21          self.maxcol = 0
22          self.minrow = 0
23          self.maxrow = 0
```

LAYER: layername
LEVEL: level_number, min_column, min_row, max_column, max_row

Table 9.1 Layer response list template.

LAYER: roads
LEVEL: 1, 0,0, 1,0
LEVEL: 2, 0 , 0, 3,1
LEVEL: 3, 0,0, 7, 3
LEVEL: 4,0,0,15,7
LEVEL: 5,0,0,31,15
LEVEL: 6,0,0,63,31
LEVEL: 7,0,0,127,63
LEVEL: 8,0,0,255,127
LAYER: imagery
LEVEL: 1,1,0,1,0
LEVEL: 2,2,1,2,1
LEVEL: 3,4,3,4,3
LEVEL: 4,8,6,8,6
LEVEL: 5,17,12,17,12
LEVEL: 6,34,24,34,24
LEVEL: 7,69,48,69,48
LEVEL: 8,139,96,139,97
LEVEL: 9,278,193,278,194
LEVEL: 10,556,387,556,388
LEVEL: 11,1112,775,1113,776
LEVEL: 12,2225,1551,2227,1553
LEVEL: 13,4451,3103,4455,3107
LEVEL: 14,8902,6207,8911,6214

Table 9.2 Example response list with two layers.

Parameter Name	Purpose	Possible Values
REQUEST	Specifies the type of request to be executed.	GETTILE TILEEXISTS GETFORMAT
LAYER	Specifies the name of the layer that will be used.	Any of the layer names listed in the layer list.
LEVEL	Specifies the zoom level number to be queried.	Any level number listed in the layer list.
ROW	Specifies the row of the tile be queried.	Any row value within the range list in the layer list.
COLUMN	Specifies the column of the tile be queried.	Any column value within the range list in the layer list.

Table 9.3 Tile request parameters.

Request types `GETTILE`, `TILEEXISTS`, and `GETFORMAT` are used to retrieve a tile image, check if the tile exists, and retrieve the format of the tiled image, respectively. An example query to retrieve a tiled image looks like the following:

```
http://www.sometileserver.com/tiles?REQUEST=
GETTILE&LAYER=imagery&LEVEL=6&ROW=24&COLUMN=34
```

Our first implementation of this scheme is in Java and uses the Java Servlet API, see Listing 9.3. The Java Servlet API provides a simple framework for creating HTTP based web services. Alternatively, our Python version of the same program uses the Web Server Gateway Interface (WSGI), see Listing 9.4. WSGI is a cross-platform interface used to support web applications in Python. WSGI provides similar functionality to the Java Servlet API.

9.4 Tile Serving Scheme with Encoded Paths

This scheme differs from the previous one in that the tile parameters are encoded in the URL path directly and not as query parameters. The template for the query method is as follows:

```
http://www.sometileserver.com/tiles/LAYERNAME/LEVEL/
ROW/COLUMN/REQUEST
```

For example, the previous tile query URL:

```
http://www.sometileserver.com/tiles?REQUEST=
GETTILE&LAYER=imagery&LEVEL=6&ROW=24&COLUMN=34
```

will now be:

```
http://www.sometileserver.com/tiles/imagery/6/24/34/
GETTILE
```

The method provides shorter query URLs, and mimics the paths of a computer's file system. If only the base path is entered, it will return a list of available layers and their tile ranges, exactly as with the previous scheme.

Listings 9.5 and 9.6 show implementations of this scheme. The method getPathInfo() is built into the Java Servlet API. It provides the path information after the address of the servlet. This is precisely the information we need to parse to get the path query values. The Python equivalent code uses the environ ['PATH_INFO'] value to obtain the URL path.

As seen from the example code these two schemes are very similar. The second scheme might be needed to provide compatibility with certain tiled image clients or with Internet caching systems that use URL patterns to determine what content to cache.

9.5 Service Metadata Alternatives

Each of the previous schemes returned text-formatted metadata about the layers available from its data store. Some users may prefer more formally defined formats. Two examples are XML and JSON. XML (Extensible Markup Language) and JSON (JavaScript Object Notation) both support structured, hierarchical storage of text information. Recall the layer list from previous section:

LAYER: roads
LEVEL: 1,0,0,1,0
LEVEL: 2,0,0,3,1
LEVEL: 3,0,0,7,3
LEVEL: 4,0,0,15,7
LEVEL: 5,0,0,31,15
LEVEL: 6,0,0,63,31
LEVEL: 7,0,0,127,63
LEVEL: 8,0,0,255,127

In XML, this might look like the following:

```
<Layer name="roads">
    <Level number="1" mincol="0" minrow="0" maxcol="1" maxrow="0" />
    <Level number="2" mincol="0" minrow="0" maxcol="3" maxrow="1" />
    <Level number="3" mincol="0" minrow="0" maxcol="7" maxrow="3" />
    <Level number="4" mincol="0" minrow="0" maxcol="15" maxrow="7" />
    <Level number="5" mincol="0" minrow="0" maxcol="31" maxrow="15" />
    <Level number="6" mincol="0" minrow="0" maxcol="63" maxrow="31" />
    <Level number="7" mincol="0" minrow="0" maxcol="127" maxrow="63" />
    <Level number="8" mincol="0" minrow="0" maxcol="255" maxrow="127" />
</Layer>
```

In JSON, it could be encoded as:

```
1  {
2      "layer": "roads",
3          "levels" : [
4              {"1" : [0,0,1,0] },
5              {"2" : [0,0,3,1] },
6              {"3" : [0,0,7,3] },
7              {"4" : [0,0,15,7] },
8              {"5" : [0,0,31,15] },
9              {"6" : [0,0,63,31] },
10             {"7" : [0,0,127,63] },
11             {"8" : [0,0,255,127] }
12         ]
13 }
```

The benefit of these two formats is that they are well supported by many applications and programming environments. Most web browsers can automatically convert the JSON text into in-memory objects. Optional encodings can easily be added to our tile server schemes. We can simply add a parameter called "FORMAT" with the possible values "plaintext", "XML", "JSON." The tile server can then use this parameter to determine which format to output as a response.

9.6 Conclusions

In this chapter, we have covered how to serve tiled images via HTTP. Chapter 13 provides a discussion and examples of advanced tile serving to support a variety of client software systems. We have not covered the important topic of securing our tile server or its services. For more information on this topic we suggest [1] for further reading.

Listing 9.3 Java servlet code for tile serving with encoded parameters.

```
1  public class TileServlet extends HttpServlet {
2
3      DataStore dataStore;
4
5      public void doGet(HttpServletRequest request, HttpServletResponse response)
            {
6          String requestType = request.getParameter("REQUEST");
7          if (requestType.equalsIgnoreCase("GETTILE") || requestType.
                equalsIgnoreCase("TILEEXISTS") || requestType.equalsIgnoreCase("
                GETFORMAT")) {
8              String layerName = request.getParameter("LAYER");
9              int level = Integer.parseInt(request.getParameter("LEVEL"));
10             long row = Long.parseLong(request.getParameter("ROW"));
11             long column = Long.parseLong(request.getParameter("COLUMN"));
12
13             if (requestType.equalsIgnoreCase("GETTILE")) {
14                 byte[] data = dataStore.getTileImage(layerName, level, row,
15                 column);
16                 if (data != null) {
17                     response.setContentType(dataStore.getTileFormat(layerName,
                            level, row, column));
18                     OutputStream os;
19                     try {
20                         os = response.getOutputStream();
21                         os.write(data);
22                         os.close();
23                     } catch (IOException e) {
24                         e.printStackTrace();
25                     }
26                 }
27             }
28             if (requestType.equalsIgnoreCase("TILEEXISTS")) {
29                 boolean val = dataStore.tileExists(layerName, level, row,
                        column);
30                 response.setContentType("text/plain");
31                 PrintWriter pw;
32                 try {
33                     pw = response.getWriter();
34                     pw.println(val);
35                     pw.close();
36                 } catch (IOException e) {
37                     e.printStackTrace();
38                 }
39                 return;
40             }
41             if (requestType.equalsIgnoreCase("GETFORMAT")) {
42                 String format = dataStore.getTileFormat(layerName, level, row,
                        column);
43                 response.setContentType("text/plain");
44                 PrintWriter pw;
45                 try {
46                     pw = response.getWriter();
47                     pw.println(format);
48                     pw.close();
49                 } catch (IOException e) {
50                     e.printStackTrace();
51                 }
52                 return;
53             }
54         }
55         // valid request type not found, send back layer list response
56         printLayerList(request, response);
57     }
58
59     private void printLayerList(HttpServletRequest request, HttpServletResponse
            response) {
60         PrintWriter pw;
61         try {
62             pw = response.getWriter();
63             String[] layers = dataStore.getLayersAvailible();
```

```
64        for (int i = 0; i &lt; layers.length; i++) {
65            pw.println("LAYER:" + layers[i]);
66            TileRange[] ranges = dataStore.getTileRanges(layers[i]);
67            for (int j = 0; j &lt; ranges.length; j++) {
68                if (ranges[j] != null) {
69                    pw.println("LEVEL:" + ranges[j].level + ","
70                    + ranges[j].mincol + "," + ranges[j].minrow
71                    + "," + ranges[j].maxcol + ","
72                    + ranges[j].maxrow);
73                }
74            }
75        }
76        pw.close();
77    } catch (IOException e) {
78        e.printStackTrace();
79    }
80    }
81 }
```

Listing 9.4 Python code for tile service with encoded parameters.

```python
import cgi                # the cgi module is only used for some query string parsing
class TileWSGIApp :

    def __init__(self , datastore ):
        self._datastore = datastore

    def doGet(self , environ , start_response ):
        status = '200 OK'
        headers = [('Content-type', 'text/plain')]

        queryParams = cgi.FieldStorage (environ['wsgi.input'], environ=environ )

        requestType = queryParams.getfirst ('REQUEST', '').lower ()
        if (requestType in ['gettile', 'tileexists', 'getformat']):
            layerName = queryParams.getfirst ('LAYER', '')
            level = int(queryParams.getfirst ('LEVEL', -1))
            row = int(queryParams.getfirst ('ROW', -1))
            column = int(queryParams.getfirst ('COLUMN', -1))

            if (requestType == 'gettile'):
                data = self._datastore.getTileImage (layerName , level , row ,
                    column )
                if (data != None):
                    headers = [('Content-type',
                                self._datastore.getTileFormat(layerName , level ,
                                                              row , column ))]

            elif(requestType == 'tileexists'):
                exists = self._datastore.tileExists (layerName , level , row ,
                    column )
                if (exists):
                    data = 'True'
                else :
                    data = 'False'

            elif(requestType == 'getformat'):
                data = '%s' % self._datastore.getTileFormat(layerName , level ,
                                                            row , column )

            else :
                data = self._getLayerList()
        else :
            data = self._getLayerList()

        start_response (status , headers )
        return [data]

    def _getLayerList(self ):
        strbuf = cStringIO.StringIO ()
        layers = self._datastore.getLayersAvailable ()
        for layer in layers :
            strbuf.write ('LAYER: %s\n' % layer )
            tileRanges = self._datastore.getTileRanges (layer )
            for tileRange in tileRanges :
                if (tileRange != None):
                    strbuf.write ('LEVEL: %s,%s,%s,%s,%s\n' %
                                  (tileRange.level , tileRange.mincol ,
                                   tileRange.minrow , tileRange.maxcol ,
                                   tileRange.maxrow ))
        return strbuf.getvalue ()
```

Listing 9.5 Java code for tile serving with path encoded parameters.

```java
public class TileServlet2 extends HttpServlet {

    DataStore dataStore;

    public void doGet(HttpServletRequest request, HttpServletResponse response) {
        // get any parts of the request path, after the servlet address, but not
        //      including query parameters
        String path = request.getPathInfo();
        if (path == null) {
            printLayerList(request, response);
            return;
        }
        if (path.startsWith("/")) {
            path = path.substring(1);
        }
        String[] queryVals = path.split("/");
        if (queryVals.length != 5) {
            printLayerList(request, response);
            return;
        }

        String layerName = queryVals[0];
        int level = Integer.parseInt(queryVals[1]);
        long row = Long.parseLong(queryVals[2]);
        long column = Long.parseLong(queryVals[3]);
        String requestType = queryVals[4];

        if (requestType.equalsIgnoreCase("GETTILE")) {
            byte[] data = dataStore.getTileImage(layerName, level, row, column);
            if (data != null) {
                response.setContentType(dataStore.getTileFormat(layerName,
                level, row,
                column));
                OutputStream os;
                try {
                    os = response.getOutputStream();
                    os.write(data);
                    os.close();
                } catch (IOException e) {
                    e.printStackTrace();
                }
            }
        }
        if (requestType.equalsIgnoreCase("TILEEXISTS")) {
            boolean val = dataStore.tileExists(layerName, level, row, column);
            response.setContentType("text/plain");
            PrintWriter pw;
            try {
                pw = response.getWriter();
                pw.println(val);
                pw.close();
            } catch (IOException e) {
                e.printStackTrace();
            }
            return;
        }
        if (requestType.equalsIgnoreCase("GETFORMAT")) {
            String format = dataStore.getTileFormat(layerName, level, row,
            column);

            response.setContentType("text/plain");
            PrintWriter pw;
            try {
                pw = response.getWriter();
                pw.println(format);
                pw.close();
```

```
66                          } catch (IOException e) {
67                              e.printStackTrace ();
68                      }
69                  return;
70              }
71
72              // valid request type not found, send back layer list response
73              printLayerList (request , response );
74      }
75
76      private void printLayerList (HttpServletRequest request , HttpServletResponse
            response ) {
77              PrintWriter pw;
78              try {
79                  pw = response . getWriter ();
80                  String [] layers = dataStore . getLayersAvailible ();
81                  for (int i = 0; i < layers . length; i++) {
82                      pw . println ("LAYER:" + layers [i ]);
83                      TileRange [] ranges = dataStore . getTileRanges (layers [i ]);
84                      for (int j = 0; j < ranges . length; j++) {
85                          if (ranges [j] != null) {
86                              pw . println ("LEVEL:" + ranges [j ]. level + ","
87                                  + ranges [j ]. mincol + "," + ranges [j ]. minrow
88                                  + "," + ranges [j ]. maxcol + ","
89                                  + ranges [j ]. maxrow );
90                          }
91                      }
92                  }
93                  pw . close ();
94              } catch (IOException e) {
95                  e . printStackTrace ();
96              }
97      }
98 }
```

Listing 9.6 Python code for tile serving with path encoded parameters.

```python
 1  class TileWSGIApp2:
 2
 3      def __init__(self, datastore):
 4          self._datastore = datastore
 5
 6      def doGet(self, environ, start_response):
 7          status = '200 OK'
 8          headers = [('Content-type', 'text/plain')]
 9
10          path = environ['PATH_INFO']
11
12          if (path == None):
13              data = self._getLayerList()
14          else:
15              if (path[0] == '/'):
16                  path = path[1:]
17
18              queryVals = path.split("/")
19
20              if (len(queryVals) != 5):
21                  data = self._getLayerList()
22              else:
23                  layerName = queryVals[0]
24                  level = int(queryVals[1])
25                  row = int(queryVals[2])
26                  column = int(queryVals[3])
27                  requestType = queryVals[4].lower()
28
29                  if (requestType == 'gettile'):
30                      data = self._datastore.getTileImage(layerName, level, row,
                            column)
31                      if (data != None):
32                          headers = [('Content-type',
33                                          self._datastore.getTileFormat(layerName,
                                          level,
34                                                                     row, column))
                                          ]
35
36                  elif(requestType == 'tileexists'):
37                      exists = self._datastore.tileExists(layerName, level, row,
                            column)
38                      if (exists):
39                          data = 'True'
40                      else:
41                          data = 'False'
42
43                  elif(requestType == 'getformat'):
44                      data = '%s' % self._datastore.getTileFormat(layerName,
                            level,
45                                                                     row, column)
46
47                  else:
48                      data = self._getLayerList()
49
50          start_response(status, headers)
51          return [data]
52
53      def _getLayerList(self):
54          strbuf = cStringIO.StringIO()
55          layers = self._datastore.getLayersAvailable()
56          for layer in layers:
57              strbuf.write('LAYER: %s\n' % layer)
58              tileRanges = self._datastore.getTileRanges(layer)
59              for tileRange in tileRanges:
60                  if (tileRange != None):
61                      strbuf.write('LEVEL: %s,%s,%s,%s,%s\n' %
```

```
62                          (tileRange.level,  tileRange.mincol,
63                           tileRange.minrow,  tileRange.maxcol,
64                           tileRange.maxrow))
65          return  strbuf.getvalue()
```

References

1. Wells, C.: Securing AJAX applications. O'Reilly (2007)

Chapter 10
Map Projections

10.1 Introduction to Datums, Coordinate Systems, and Projections

Any system that uses maps must take into account three properties: datums, coordinate systems, and projections. These properties specify our model of the Earth and the way in which we specify locations upon it. In the world of geodesy there are a number of different options for each of these properties with no one "best fit" choice for all applications. While it is often not necessary to use more than one datum or coordinate system, it is important to understand them in case there is a need for interoperability with another system or data source which uses a different datum or coordinate system. As such, we will provide a general introduction to these two properties.

This chapter will focus primarily on map projections, the means by which the three-dimensional Earth is represented in two-dimensional map images. While we will provide a basic introduction to map projections, the focus of this chapter will be on converting between different map projections and how projections are used in a tile-based mapping system.

10.1.1 The Shape of the Earth

Before we can discuss referencing locations on the surface of the Earth or projecting the surface onto a flat plane, we must define the shape of the Earth. The technical term for the true shape of the Earth is the geoid. Formally, the geoid is a surface where gravity's strength is equal at mean sea level. For our purposes it is only important to note that a geoid is not a flat surface but has variations over the surface of the

J.T. Sample and E. Ioup, *Tile-Based Geospatial Information Systems:*
Principles and Practices, DOI 10.1007/978-1-4419-7631-4_10,
© Springer Science+Business Media, LLC 2010

Earth. Figure 10.1[1] shows gravity variations over the Earth's surface. Figure 10.2[2] shows the variations between the geoid and the most common approximation of the shape of the Earth. The earliest approximations of the geoid used a sphere as the shape of the Earth. In fact, a sphere is often used today to approximate the shape of the Earth. The errors of a spherical approximation are not visible at whole Earth scales.

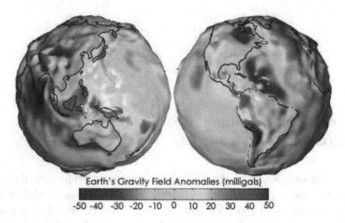

Fig. 10.1 Gravity variations over the surface of the Earth.

A sphere is not the best approximation of the Earth because of flattening at the poles. Instead, the best approximation is the elliptical analog to a sphere, called either an ellipsoid or spheroid (see Figure 10.3). A number of different ellipsoid approximations are currently in use. Different ellipsoids may provide better approximations for specific areas of the Earth. There are also a number of ellipsoids which have been superceded but still can be found in older maps or measurements.

10.1.2 Datums

Individual ellipsoids are often referred to in the context of a datum. A datum is an ellipsoid along with an origin point from which locations are referenced (see Figure 10.4). Datums may be global and cover the entire Earth, or they may be local and designed for mapping over a very small area. We are most interested in global datums for the purpose of tile-based mapping; however, it is possible that a local datum will be encountered in source data. Converting between datums is possible,

---

[1] Courtesy of NASA. http://earthobservatory.nasa.gov/Features/GRACE/page3.php

[2] http://commons.wikimedia.org/wiki/File:Geoid_height_red_blue.png

Deviation of the Geoid from the idealized figure of the Earth
(difference between the EGM96 geoid and the WGS84 reference ellipsoid)

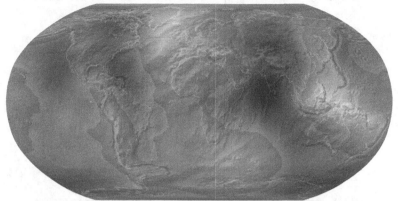

Red areas are above the idealized ellipsoid; blue areas are below.

| -107.0 m | 0 m | +85.4 m |

Fig. 10.2 Deviation of the geoid from the shape of the most commonly used ellipsoid approximation.

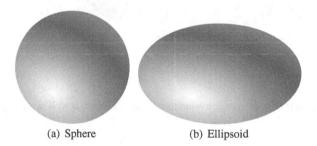

(a) Sphere   (b) Ellipsoid

Fig. 10.3 The sphere and ellipsoid are commonly used as approximations for the shape of the Earth. The above ellipsoid has an exaggerated flattening to highlight its shape. Earth approximating ellipsoids visually resemble a sphere because they have only a small amount of flattening.

and the references at the end of the chapter are a good starting point if information on how to perform the conversions is necessary.

The most commonly used datum is the World Geodetic System 1984, commonly abbreviated WGS84. The WGS84 datum has a corresponding ellipsoid that provides a good approximation for the entire Earth. While locally oriented ellipsoids will provide better approximations for small areas, the WGS84 ellipsoid was designed to be a "best fit" for the entire Earth. The origin for the WGS84 datum is the center of the Earth, rather than a point on the surface as is common for many local datums. This geocentric origin is necessary for satellite systems to use the datum.

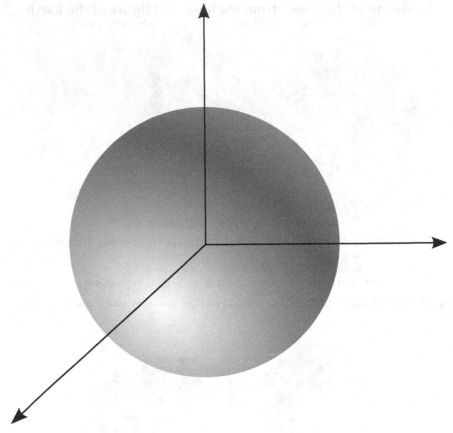

Fig. 10.4 A datum is both an approximation for the Earth's shape as well as an origin.

As stated above, WGS84 is by far the most commonly used datum for geospatial data. Most source data used in a tile-based mapping system will use the WGS84 datum. Some exceptions are the NAD27 or WGS72 datums, both obsolete datums once in use in the United States. The only reason to use a local datum is when a mapping application requires greater local geoid accuracy than the WGS84 ellipsoid can provide. This level of accuracy is rarely needed in a tile-based mapping system. On the other hand, the WGS84 datum provides the most compatibility with other software applications and data sources. Interoperability is an increasingly important component of mapping systems. Thus, in most cases, a tile-based mapping system should always use the WGS84 datum for native backend-data. The added benefit of using WGS84 is that it is so common that most data will already be in this datum with no additional work [2].

10.1.3 Coordinate Systems

Once a datum is selected to represent the shape the Earth, a coordinate system must be defined to specify locations on its surface. There are two coordinate system types in use today. The first is the commonly used geographic coordinate system which uses latitude and longitude to specify a location. Geographic coordinates are a basic angular coordinate system, either over a sphere or ellipsoid depending on which approximation is used. The latitude is the angle north or south from the equator in the range -90 to 90 degrees. The longitude is the angle east or west from the Central (Greenwich) meridian in the range -180 to 180 degrees.

An alternate way of specifying locations on the surface of the Earth is with rectangular coordinates. Rectangular coordinates are a Cartesian coordinate system with X and Y coordinates to represent horizontal and vertical position, respectively. Because a Cartesian coordinate system is used for a two-dimensional grid, rectangular coordinates are only used on flat maps. Usually, the coordinates are specified using meters. As a result, they are useful for performing calculations, such as distance measurements, where the angular geographic coordinates would be cumbersome. Rectangular coordinates are rarely used for maps of small scale, i.e. covering a wide area of the Earth. The distortions associated with the flattening of the Earth become so large that calculations with rectangular coordinates would be of no use. Thus, rectangular coordinates are limited to maps of smaller areas such as topographic maps or high-resolution aerial imagery [6].

10.2 Map Projections

As discussed earlier, the Earth is a geoid which is either approximated as a sphere or ellipsoid. Using two-dimensional maps of the Earth, such as a paper map or satellite image, requires transforming the surface of the Earth to a plane, called a projection. Unfortunately, transforming the surface of a sphere onto a plane causes distortion. It is mathematically impossible to design a projection which does not cause some type of distortion. As a result, there are a number of different types of projections in common usage which limit one type of distortion in exchange for increasing others. The discussion of specific map projections in this chapter will be limited to common projections most likely to be encountered by the implementor of a tile-based mapping system.

Whenever the surface of the Earth is projected onto a plane there is distortion. A number of different types of distortion can occur depending on which map projection is used, including distortions of area, shape, distance, direction, and angle. Individual map projections often reduce or eliminate one of these distortion types. For example, Albers' equal-area projection removes distortion of area on the map at the cost of increasing distortion in other areas [7].

For tile-based mapping there are three important types of maps which are important to recognize. The first type is the equidistant map which most commonly preserves the length of the meridians. Each line of longitude is of the same length in an equidistant map, so that distances measured on a line of longitude are the same as on the globe. The second type of map is equal-area which preserves the area occupied by a feature. The third map type is conformal which preserves the shape of features. An additional property of conformal projections is that any straight line on the map forms an angle of constant bearing with each line of longitude. This latter property is useful for navigation and led to the historic popularity of conformal projections, such as the Mercator projection.

A projection may be created by directly mapping the globe to a planar surface. However, it is also possible to perform the trasformation by first projecting the globe to an intermediate shape. The two most common intermediate shapes are a cylinder and a cone. Both the cylinder and the cone may be transformed into a plane with no distortion. While not all projections are formed in this manner, many of the most commonly used projections are either cylindrical, conic, or planar, as shown in Figure 10.5. Distortion is always minimal at the location where the shape touches the globe, such as the equator for many cylindrical projections [6].

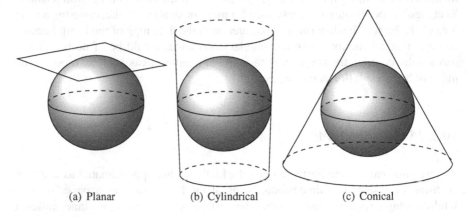

(a) Planar   (b) Cylindrical   (c) Conical

Fig. 10.5 Projections are created by projecting Earth's surface onto a surface that may be transformed into a plane.

10.2.1 Different Map Projections

Theoretically, there are an infinite number of different map projections. In practice, there are merely dozens in common usage on paper maps. For the purpose of tile-based mapping, there are only a few map projections which are important to discuss. Tile-based mapping systems, or any computer mapping system, must focus heavily on interoperability. Thus, the projections used in these systems tend to be limited to a few standard projections. In contrast, paper maps do not need to interoperate with any other data or system and may select the projection with the best utility for the map's specific view of the Earth.

10.2.2 Cylindrical Equidistant Projection

The first important projection for tile-based mapping is the cylindrical equidistant projection. This projection is sometimes called the Cylindrical Equirectangular, Plate Carrée, Simple Cylindrical, WGS84 Geodetic, or WGS84 Lat/Lon projection [7]. The cylindrical equidistant projection does not scale the meridians of the original globe. Thus, distances are not distorted north-to-south. However, each line of latitude is stretched to the same length as the equator, providing significant stretching east-to-west. The distortion in this projection ranges from 0 at the equator to infinity at the poles. Additionally, neither area, shape, nor bearing is preserved on the map. Figure 10.6 is an example of a geodetic map of the world.

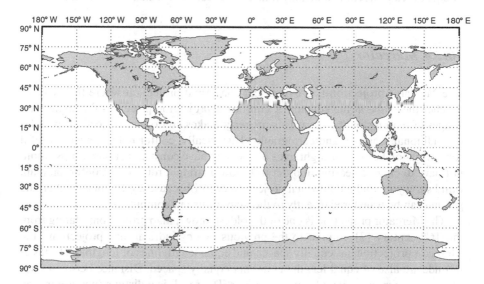

Fig. 10.6 Geodetic projected map of the world.

The benefit of this map projection is the ease of construction, especially for computer mapping systems. One degree of latitude and one degree of longitude are the same length in any area of the map. The same is not true of other projections (see Mercator projection below). As a result, the resulting map forms a simple Cartesian coordinate system of the world, centered at longitude and latitude (0,0). Another useful consequence is that each pixel in a map image represents the same distance in both the east-west and north-south directions, simplifying calculations performed on map images.

10.2.3 Cylindrical Equal-Area Projection

The cylindrical equidistant projection is simple and easy to use; however, it offers little beneficial map properties other than simplicity. The parallels are stretched by a value of $\sec \lambda$, where λ is the latitude. By scaling the projection by a factor of $\cos \lambda$, we can create the cylindrical equal-area projection. The cylindrical equal-area projection preserves the area of features but not the distances between them or angles on the map.

The cylindrical equal-area projection is a rarely used projection, in both printed maps and computer map images. However, it is useful when used with data in the cylindrical equidistant projection. Maps at large scale can be reprojected to equal-area by multiplying by the cosine of the average latitude of the map. Reprojecting to equal-area projection at display time can reduce visible distortions of features.

10.2.4 Mercator

The Mercator projection is a cylindrical conformal projection where the cylinder touches the globe at the equator. A conformal projection preserves the angles and shapes of features on the map. This property makes conformal projections popular in traditional mapping and cartography. Historically, conformal maps were useful for navigation because straight lines on the map have constant bearing. Ships would not have to change bearing in order to follow a straight route on a conformal map. Land surveyors find a conformal projection useful because angles measured on the ground can be transfered directly to the map for use in computation.

The Mercator projection is one of the oldest known map projections, dating from the 16th century [7]. Similar to the previous two cylindrical map projections, the Mercator projection has equally spaced and equal length lines of longitude. Lines of latitude are also of equal length but are not equally spaced. Distance between lines of latitude increases away from the equator. The result is a distinct increase in size of features towards the poles. The features themselves are the same shape as on the globe. Figure 10.7 is an example of a Mercator map of the world.

10.2.5 Universal Transverse Mercator

The Transverse Mercator projection is a cylindrical projection similar to the standard Mercator projection. However, rather than deriving the projection from a cylin-

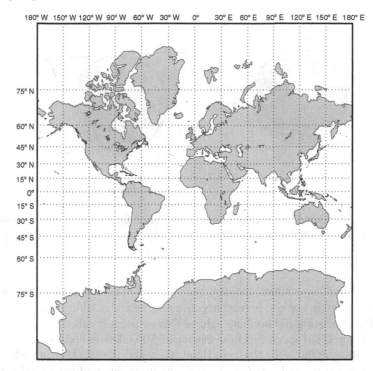

Fig. 10.7 Mercator projected map of the world.

der oriented vertically pole to pole, the Transverse Mercator projection is based on a cylinder oriented horizontally parallel to the equator, as shown in Figure 10.8. Lines of latitude and longitude are no longer straight in the Transverse Mercator projection. Whereas the Mercator projection distorts features farther north or south, the Transverse Mercator projection distorts features lying farther east or west from the central meridian. The projection can be made using any line of longitude as the central meridian and provides little distortion to areas within a short range. As a result, the Transverse Mercator projection is often used for maps of small scale for use in land surveying. Transverse Mercator maps primarily use rectangular coordinates (meters) to simplify ground calculations over small areas.

Universal Transverse Mercator (UTM) is a standard coordinate and projection system commonly used for imagery, topographic maps, and other low scale geospatial data. The UTM system covers the entire world from 80°S to 84°N. Each hemisphere is split into 60 zones 6° wide. Zones are number 1 to 60 from west to east starting at 180°W. Zones are identified by their zone number and hemisphere (North or South).

Each UTM zone has its own specific projection and coordinate system. The central longitude of each zone is used as the central meridian for a Transverse Mercator projection. UTM uses a rectangular coordinate system with the origin at the intersection of the central meridian and the equator. Locations are identified by their

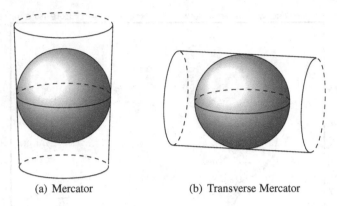

(a) Mercator   (b) Transverse Mercator

Fig. 10.8 Mercator and Transverse Mercator both project the Earth's surface onto a cylinder. The difference between the two is the orientation of the cylinder.

easting and northing values , the distance east and north in meters from the zone origin. However, in order to prevent negative easting and northing values a constant is added to all easting and northing values. This constant is called a false easting or false northing. UTM specifies a false easting of 500,000m in both hemispheres and a false northing of 0m in the Northern Hemisphere and 10,000,000m in the Southern Hemisphere. For example, in Zone 1 South, the intersection of the central meridian (174°W) and the equator is given the coordinate 500000E, 10000000N. Figure 10.9 shows the coordinates of two points before and after the conversion to false northing and easting.

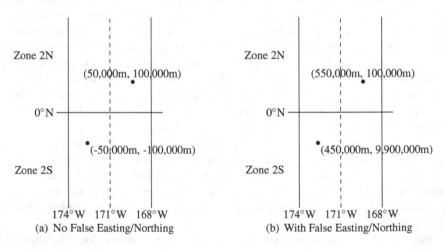

Fig. 10.9 False northing and easting values are simply an artificial translation of the coordinates to ensure they are non-negative.

Since UTM zones are only 6° wide, the amount of spatial distortion on any UTM map is small. As a result, UTM is used frequently in satellite and aerial imagery. Source data used in tiled map systems will often be projected using UTM. For example, USGS produced Digital Orthophoto Quadrangle (DOQQ) aerial imagery is distributed using UTM.

10.3 Point Reprojection

Often, it is necessary to convert between projections in the process of creating a tiled-mapping system. Source data will often be distributed using a projection such as UTM while data in the tiled-mapping system will be stored natively in a global projection such as Lat/Lon or Mercator. We will present some basic reprojection formulae and techniques and demonstrate the process of converting from UTM to Lat/Lon.

Before covering techniques for reprojecting images, we must explain how to reproject individual points. We will present the forumlas here along with some example code. The formulae and a more thorough explanation can be found in Snyder's "Map Projections: A Working Manual" [7].

Below are the formulae for converting a Geodetic coordinate into a UTM coordinate:

$$x = X_0 + k_0 N \left[A + (1 - T + C) \frac{A^3}{6} + (5 - 18T + T^2 + 72C - 58e'^2) \frac{A^5}{120} \right] \tag{10.1}$$

$$y = Y_0 + k_0 \left\{ M - + N \tan \phi \left[\frac{A^2}{2} + (5 - T + 9C + 4C^2) \frac{A^4}{24} + (61 - 58T + T^2 + 600C - 330e'^2) \frac{A^6}{720} \right] \right\} \tag{10.2}$$

where

$$e = \sqrt{1 - \frac{b^2}{a^2}} \tag{10.3}$$

$$e'^2 = \frac{e^2}{(1 - e^2)} \tag{10.4}$$

$$N = \sqrt{\frac{a}{1 - e^2 \sin^2 \phi}} \tag{10.5}$$

$$T = \tan^2 \phi \tag{10.6}$$

$$C = e'^2 \cos^2 \phi \tag{10.7}$$

$$A = (\lambda - \lambda_0) \cos \phi \tag{10.8}$$

$$M = a[(1 - e^2/4 - 3e^4/64 - 5e^6/256 - \ldots)\phi - (3e^2/8 + 3e^4/32 + 45e^6/1024 + \ldots) \sin^2 \phi$$
$$+ (15e^4/256 + 45e^6/1024 + \ldots) \sin^4 \phi - (35e^6/3072 + \ldots) \sin^6 \phi + \ldots] \tag{10.9}$$

and

ϕ = latitude (radians)

λ = longitude (radians)

λ_0 = central longitude of the UTM zone (radians)

$a = 6,378,137$

$b = 6,356,752.3$

$k_0 = 0.9996$

$X_0 = 500,000$

$$Y_0 = \begin{cases} 0 & \phi \geq 0 \\ 10,000,000 & \phi < 0 \end{cases}$$

Code for transforming a point from Geodetic to UTM is given in Listing 10.2.

The inverse formulae for converting from UTM to Geodetic coordinates are below. Variable definitions are equivalent to the forward conversion formulae unless otherwise noted.

$$\phi = \phi_1 - \left(N_1 \frac{\tan\phi_1}{R_1}\right)\left[\frac{D^2}{2} - (5 + 3T_1 + 10C_1 - 4C^2 - 9e'^2)\frac{D^4}{24}\right.$$

$$\left. + (61 + 90T_1 + 298C_1 + 45T_1^2 - 252e'^2 - 3C_1^2)\frac{D^6}{720}\right] \tag{10.10}$$

$$\lambda = \lambda_0 + \left[D - (1 + 2T_1 + C_1)\frac{D^3}{6} + (5 - 2C_1 + 28T_1 - 3C_1^2 + 8e'^2 + 24T_1^2)\frac{D^5}{120}\right]\frac{1}{\cos\phi_1} \tag{10.11}$$

$$\phi_1 = \mu + \left(\frac{3e_1}{2} - \frac{27e_1^3}{32} + \ldots\right)\sin 2\mu + \left(\frac{21e_1^2}{16} - \frac{55e_1^4}{32} + \ldots\right)\sin 4\mu \tag{10.12}$$

$$+ \left(\frac{151e_1^3}{96} - \ldots\right)\sin 6\mu + \left(\frac{1097e_1^4}{512} - \ldots\right)\sin 8\mu + \ldots \tag{10.13}$$

where

$$e_1 = \frac{1 - \sqrt{1 - e^2}}{1 + \sqrt{1 - e^2}} \tag{10.14}$$

$$\mu = \frac{M}{a\left(1 - \frac{e^2}{4} - \frac{3e^4}{64} - \frac{5e^6}{256} - \ldots\right)} \tag{10.15}$$

$$M = \frac{y - Y_0}{k_0} \tag{10.16}$$

$$e'^2 = \frac{e^2}{(1 - e^2)} \tag{10.17}$$

$$C_1 = e'^2 \cos^2\phi_1 \tag{10.18}$$

$$T_1 = \tan^2\phi_1 \tag{10.19}$$

$$N_1 = \sqrt{\frac{a}{1 - e^2\sin^2\phi_1}} \tag{10.20}$$

$$R_1 = \frac{a(1 - e^2)}{(1 - e^2\sin e^2\phi_1)^{\frac{3}{2}}} \tag{10.21}$$

$$D = \frac{x - X_0}{N_1 k_0} \tag{10.22}$$

Both the forward and inverse conversion formulae are based on series expansion approximations. The expansions converge quickly and are accurate to about a centimeter given the 6° width of a UTM zone. It is important to note the extensive floating point math in these formulae. Using double-precision floating point math with these formulae is quite expensive. Performing a single Geodetic to UTM conversion will appear instantaneous on modern day computers. However, when scaled to thousands, or even millions, of conversions, the costs become noticeable.

10.4 Map Reprojection

The previous section dealt only with converting single locations from one projection to another. While important, individual point reprojection is not the end goal for tiled-mapping. What we really must accomplish is reprojection of entire map images. Reprojecting an entire map image is not a simple matter of applying formulae. We must weigh a number of considerations when determining what techniques to use for image reprojection.

Map images are often large and reprojecting them can be computationally costly. Combined with the large numbers of images in many map image datasets, the issue of algorithm efficiency becomes important. On the other hand, we have the problem of geospatial accuracy. First, we must ensure that a feature in the map image has coordinates representing where it actually is on the globe. The UTM/Geodetic point reprojection formulae presented above do have high accuracy given the 6° width of UTM zones . However, there are a number of different method of doing image reprojection, some of which sacrifice accuracy in exchange for improved performance. In many cases, it is acceptable to sacrifice accuracy in a tiled mapping system. Centimeter accuracy is not necessary in an online mapping system for the layperson. There is another type of distortion which can be highly damaging in a tiled-mapping system: border distortion, the discontinuities between images, is unacceptable in a tiled-mapping system. No level of user will accept discontinuities between tiles. As such, it is imperative that our reprojection technique eliminate visible discontinuities at the edges of images.

10.4.1 Affine Transforms

Affine transforms provide one possible means of reprojecting map images [2]. An affine transform is a linear transformation plus a translation.

$$u = a_0 + a_1 x + a_2 y$$
$$v = b_0 + b_1 x + b_2 y$$

The linear transform in an affine transform may be a combination of rotation, scaling, or shear. Affine transforms preserve collinearity of points. Straight lines before the transform are also straight after the transform. Affine transforms may be represented as a matrix and are computationally simple to apply to an entire image because the transform is linear. Many image libraries provide the functionality of applying an arbitrary affine transform to an image (e.g., Java, Python Imaging Library).

To reproject an image from UTM to Geodetic using an affine transform, we must determine the parameters of the transform and then apply the transform to the image. Determining the parameters requires solving a simple matrix equation:

$$\begin{pmatrix} 0 & w-1 & w-1 \\ 0 & 0 & h-1 \\ 1 & 1 & 1 \end{pmatrix} = \begin{pmatrix} a & b & c \\ d & e & f \\ 1 & 1 & 1 \end{pmatrix} \begin{pmatrix} x_1 & x_2 & x_3 \\ y_1 & y_2 & y_3 \\ 1 & 1 & 1 \end{pmatrix}$$

The variables w and h are the width and height of the reprojected image. The (x_1, y_1), (x_2, y_2), and (x_3, y_3) are the north-west, north-east, and south-east corner pixel coordinates of the Lat/Lon quad inside the UTM image. These are the pixel locations of the corner points of the Lat/Lon image individually reprojected into the UTM image (using the point reprojection formular using Equations 10.1 and 10.2). The affine transform is the solution defined by the variables a, b, c, d, e, and f. When applied to the UTM image, the three image corner points used to solve the equation will map perfectly to the Lat/Lon image. The final corner will be slightly displaced from the corner in the final image, but the difference will be small in a high resolution image.

The computational efficiency of the affine transform, combined with its ease of use in most programming environments, makes it a good candidate as a reprojection algorithm. For individual map images it is a good method of reprojection, especially if processing power is minimal and measurement distortion is not an issue.

The performance of affine transform reprojection makes it appear to be a great technique for tiled-mapping where large datasets demand efficient algorithms. However, approximating the reprojection with a linear transform causes significant border distortion. As seen in the point reprojection formulae (Equations 10.10 and 10.11), the UTM to Geodetic reprojection is non-linear. Visually, this means that the UTM grid lines and image will become curved. However, the affine transform's collinearity property ensures that they remain straight. Thus, the resulting image is not on a Geodetic grid but slightly distorted. Additionally, each reprojected image uses a unique affine transform best approximating the specific reprojection of that image. The different transforms will cause each image to be reprojected differently. These combined error sources result in significant discontinuities between reprojected images. At high resolution discontinuities in features such as roads in a tiled-mapping system are unacceptable for users and must be avoided. Thus, the affine transform does not provide an acceptable reprojection solution [3].

10.4.2 Interpolation

The alternatives to affine transform reprojection of images are techniques that combine point reprojection with interpolation. The interpolation algorithms can be confusing, so we will explain them below.

Interpolation is a technique to approximate data values that lie within some set of known data values. Linear interpolation is a commonly used and simple interpolation method. It is based on the assumption that the underlying function modeling the data values is linear. As a simple example, if we have known data values $(2, 10)$ and $(3, 20)$, then using linear interpolation, we can approximate values $(2.5, 15)$, $(2.7, 17)$, $(2.1, 11)$, etc.

But how do we interpolate values of image pixels? Here, we have pixel location represented by a two-dimensional coordinate and color value. Rather than interpolate between two pixels lying on a straight line, we interpolate between four pixels forming a square. Bilinear interpolation provides a means of determining the value (e.g., color) of any pixel lying within any four pixels with known values.

$$f(x,y) = f(0,0)(1-x)(1-y) + f(1,0)x(1-y) + f(0,1)(1-x)y + f(1,1)xy$$

The four known pixels are normalized to coordinates $(0,0)$, $(0,1)$, $(1,0)$, and $(1,1)$. The value of a pixel is represented by the function $f(x,y)$. As seen in the formula, the value of the internal pixel is weighted more heavily towards the pixels it is closest to. For RGB color pixels, the bilinear interpolation must be performed three times, one for each color component. A visualization of bilinear interpolation is shown in Figure 10.10.

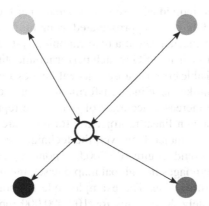

Fig. 10.10 Bilinear interpolation between four points to calculate the value of the interior point.

10.4.3 Point-wise Reprojection

The point reprojection formulae presented earlier imply a simple method for reprojecting an entire map image: take every point in the UTM image and reproject it to a
Geodetic image. Such an algorithm would work, except that a pixel in the UTM image is highly unlikely to be reprojected exactly onto a pixel in the Geodetic image.
Instead, it will map to a point between pixels. One could devise a method of reprojecting UTM points and then performing interpolation to determine values of the
Geodetic pixels. However, this would become needlessly complicated and probably
lead to a loss of accuracy.

There is a better way of performing point-wise reprojection diagrammed in Figure 10.11. First, start by creating an empty Geodetic image covering the same geographic area as the UTM image. Use the UTM to Geodetic formulae to determine
the Geodetic coordinates of the corners of the UTM image. The Geodetic image is
the bounding rectangle around those four corners. Then for each pixel in the Geodetic image, reproject the coordinates back into UTM. In general, the UTM coordinates will not correspond to one pixel in the UTM image. Instead, it will lie inside
a square bounded by four pixels. The color value for this UTM point can be calculated using bilinear interpolation using the surrounding four pixels' color values.
This color value is also the color value for the pixel in the Geodetic image. Fill in
the color and proceed to the next pixel.

Point-wise reprojection has the benefit of being accurate, both geospatially and
with border alignment. The reprojection accuracy is the same as with the point reprojection formulae with the additional small error caused by the bilinear interpolation for pixel colors. We can assume color interpolation error is minimal because
images usually do not have a lot of high frequencies in the color components. The
color variation between individual pixels is so small that the underlying function
is essentially linear and thus well approximated using bilinear interpolation. Additionally, the distribution mechanism of a tiled-mapping system will often be a lossy
compression algorithm such as JPEG, which performs smoothing on the color. Border alignment has no visible error because adjacent images are reprojected using the
same transformation, unlike the affine transformation reprojection.

The flip side of the increased accuracy of point-wise reprojection is increased
computational cost. The non-linear reprojection formulae are reasonably fast when
run once on a modern computer. However, the technique scales poorly. A 10,000
by 10,000 pixel image would require 100,000,000 point reprojections. When that is
scaled to the number of images in a global map dataset, the time required to reproject the data becomes untenable. The example code for point reprojection given
above takes approximately 20 minutes for 100,000,000 reprojections. Compiled
code would be faster but not fast enough. What we really need is an improved
algorithm [3].

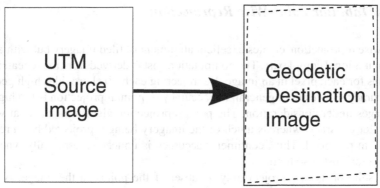

(a) Calculate the geographic area quadrilateral of the Geodetic image from the UTM source image. Create the Geodetic destination image using the bounding box of the Geodetic area quadrilateral.

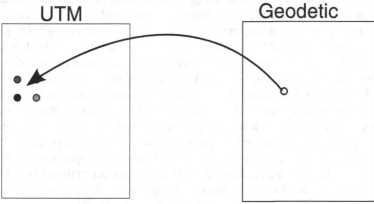

(b) Convert the coordinates of a pixel in the Geodetic image to UTM coordinates. The new UTM point will lie between four pixels in the UTM image.

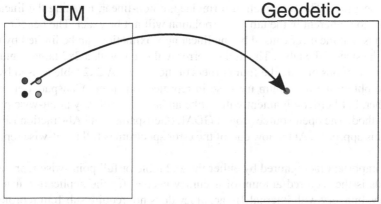

(c) Perform bilinear interpolation using the four UTM pixels to determine the color of the original Geodetic pixel.

Fig. 10.11 Point-wise reprojection of a map. The steps in figure (b) and (c) will be performed for each pixel in the Geodetic image.

10.4.4 Tablular Point-Wise Reprojection

Point-wise reprojection creates excellent alignment of tiled imagery but with a cost of computational complexity. The computation cost is derived from the repeated calculations for each pixel in an image. Reprojecting each pixel provides high geospatial accuracy. However, in general, the accuracy of point reprojection is too high for the images under consideration. The point reprojection algorithms have, at worst, centimeter accuracy, whereas much of the imagery being reprojected has a resolution of 1m per pixel. Thus, centimeter accuracy is unneeded, especially when the computation cost is so high.

Our solution is to reproject only a subset of the points in the image. A table is generated by subsampling the pixels in the Geodetic image. Thus, instead of a 10,000 by 10,000 Geodetic image, we may have a 100 by 100 table covering the same geographic area. Each pixel in the table is projected from Geodetic to UTM. (Remember, in order to convert an image from UTM to Geodetic, the coordinates of each target pixel in the Geodetic image are converted to UTM coordinates so that the target pixel's color may be calculated from the surrounding UTM pixels). The size of the table should be a divisor of the size of the desired Geodetic image and also contain the four corners of the Geodetic image to simplify the algorithms using the table. Once the table is created, it is used in the reprojection of the entire image. To reproject a Geodetic pixel we find the nearest pixels in the table and perform a bilinear interpolation to calculate each UTM component. Reprojecting the entire image requires only linear operations rather than the non-linear reprojection formulae. Figure 10.12 demonstrates the process of table-based reprojection.

Of course, there is the important question of how tabular reprojection will affect the accuracy of the reprojected image. First, there is no effect on the quality of image tile alignment. The borders of two adjacent images will still be projected using the same method, providing visually perfect alignment. Border alignment is important because discontinuities in a tiled-map system are unacceptable to users.

Geospatial accuracy is also important. The tabular point-wise reprojection will reduce accuracy. The reprojection formulae are non-linear, meaning the linear approximation inherent in the table interpolation will not be exact. The benefit of this method is that the error caused by the linear approximation can be limited by modifying the size of the table. The highest error will occur with a 2x2 table containing true reprojections of only the four corners of the image. A 2x2 table is used by systems to obtain the maximum increase in reprojection speed. Geospatial error will be higher, but border alignment will not be an issue as with any point-wise projection method. The open-source project GDAL (Geospatial Data Abstraction Library) takes this approach. At the low end of the error spectrum is full point-wise reprojection.

An important fact ignored by either the 2x2 table or full point-wise reprojection methods is the required amount of accuracy needed for the application using the imagery. Imagery with internal 1m accuracy does not require sub 1cm reprojection accuracy. It is a waste of resources to reproject an image with more accuracy than is internal to the image. A better method would be to tailor the table resolution to the

Geodetic

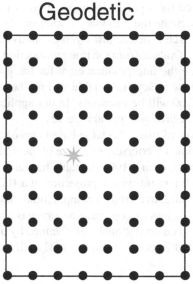

(a) Create the table by performing a UTM reprojection on a subset of the points in the Geodetic image. For other Geodetic points that must be converted, locate their positions in the table.

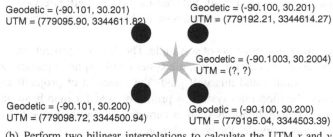

Geodetic = (-90.101, 30.201)
UTM = (779095.90, 3344611.82)

Geodetic = (-90.100, 30.201)
UTM = (779192.21, 3344614.27)

Geodetic = (-90.1003, 30.2004)
UTM = (?, ?)

Geodetic = (-90.101, 30.200)
UTM = (779098.72, 3344500.94)

Geodetic = (-90.100, 30.200)
UTM = (779195.04, 3344503.39)

(b) Perform two bilinear interpolations to calculate the UTM x and y coordinates of the target point.

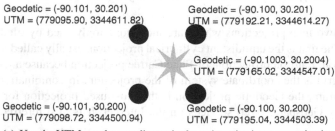

Geodetic = (-90.101, 30.201)
UTM = (779095.90, 3344611.82)

Geodetic = (-90.100, 30.201)
UTM = (779192.21, 3344614.27)

Geodetic = (-90.1003, 30.2004)
UTM = (779165.02, 3344547.01)

Geodetic = (-90.101, 30.200)
UTM = (779098.72, 3344500.94)

Geodetic = (-90.100, 30.200)
UTM = (779195.04, 3344503.39)

(c) Use the UTM x and y coordinates in the point-wise image reprojection algorithm.

Fig. 10.12 Table-based reprojection of points in an image.

level of accuracy required for a particular application or dataset. A simple binary search can be used to calculate the necessary resolution. There is no need to create the entire table in the search process, just a representative set of points to test the accuracy. A comparison of values from the true reprojection formulae and the linear interpolation will provide the interpolation error for the given table resolution. In most cases the size of an accuracy-tuned table will not be large. It will be rare that a table larger than 100x100 will be necessary. In our applications, a 16x16 table is adequate to provide our desired geospatial accuracy.

The computational cost of using the table-based approach is much smaller than full point-wise reprojection. Of course, as the size of the table increases, so does the computation cost. However, even a 100x100 table has far fewer non-linear computations in comparison to full point-wise reprojection of a 10000x10000 pixel image. The 2x2 table is highly attractive from a computational cost perspective, but it is important to ensure that the loss of geospatial accuracy is acceptable in the final application. The reduction in computational cost obtained by using a 2x2 table instead of a 16x16 table is limited in comparison to the significant reduction in accuracy [4].

10.5 Map Projections for Tiled Imagery

Once we are able to move between projections, we must decide which Coordinate Reference System (CRS) to use when storing the tiled map imagery. The best datum to use will usually be WGS84. A global tiled-map system will rarely be useful if it is in another datum. WGS84 is the most interoperable datum and the best fit for global datasets.

Choosing a map projection is not so simple. The different projections offer different useful properties. In certain cases, the system will require a particular property, equal area for example, and this will guide the selection of projection. However, we will assume the tiled-map system is primarily used for map browsing and not advanced geospatial applications such as cartography, navigation, etc. In this case, no single projection provides the best solution.

10.5.1 Storing Tiles in the Geodetic Projection

There are two map projections which are most commonly used by tiled-mapping systems . The first is the equidistant cylindrical projection, usually called the Geodetic projection, Lat/Lon projection, or Plate Carrée projection because a latitude and longitude grid is the coordinate system of the projection. In combination with the WGS84 datum, the Geodetic projection is the most used projection for map data provided by Geospatial Web services on the Internet. It is also referred to by its

EPSG code, 4326. The European Petroleum Survey Group has codes for most CRS which has become a standard means of identifying them.

The Geodetic projection is used for Web mapping application so often because it is simple. The coordinate system is latitude and longitude, and the projection forms a perfect grid with those coordinates. Images in geodetic projection are easily overlaid on a globe, such as in 3D map clients such as Google Earth or NASA World Wind. Also, mathematical operations using coordinates, common in map image manipulation, become much simpler with the Geodetic projection. Each pixel in a map image with Geodetic projection has the same dimensions in degrees.

The downside of the Geodetic projection is the distortion. Area, shape, and latitude distances are distorted in the Geodetic projection. These distortions are especially apparent close to the poles. Example distortions are circular domes which become ovals and streets which no longer intersect at right angles. While any map projection creates distortion, these types of distortions are readily apparent to users, especially on a tiled-mapping system designed to display maps at large scale (zoomed in).

As mentioned above, 3D map clients such as Google Earth and NASA World Wind use the Geodetic projection, as well as the online map tool Mapquest. GIS clients understand the Geodetic map projection. WMS clients primarily function using the Geodetic projection.

10.5.2 Storing Tiles in the Mercator Projection

A common alternative to storing tiles in the Geodetic projection is to store them using the Mercator projection . The Mercator is a cylindrical conformal projection, meaning that it preserves angles and shapes on the map. It is also a good projection for global data; most of the globe is clearly visible on a small scale map. While most interoperable geospatial Web services do not use the Mercator projection for their data, it is used in some common tiled-mapping systems on the Web. Google Maps, Yahoo Maps, Microsoft Bing Maps, and Open Street Map all store their tiled imagery in the Mercator projection.

These Web mapping sites use the Mercator projection primarily because of its angle preservation properties. The primary use of these systems is to view map images at a large scale. Users zoom into the map to retrieve street directions or search for locations. The Mercator projection prevents visible distortions in these large scale maps. Road intersections retain the same angles as on the globe. In the Geodetic projection this is not the case. Roads in areas nearer to the poles, such as Finland, intersect at non-right angles. Shapes are also preserved in the Mercator projection. Circular buildings do not become stretched. The primary distortion caused by the Mercator projection, size distortion, is not visible in these large scale maps. At the local level, sizes are uniformly increased in size around the poles, so that the distortions are not visible.

The Mercator projection has downsides for use as a projection for native tile storage. The primary problem for the Mercator projection is the issue of interoperability. Clients which access map services, especially WMS clients, usually support the Geodetic projection. Support for the Mercator projection in non-tile clients is limited. Thus, if the tiled-map system will be a backend to a WMS, using the Mercator projection will limit its use by many clients. Often tiled map data will be used as an overlay or background to other map data. Most of this map data is in the Geodetic projection, and combining with Mercator projected data will be difficult. A good example of this problem is Google Earth. It expects data to be in Geodetic projection. Using Mercator projected data as an overlay in Google Earth will cause problems. Web mapping systems such as Google Maps are not concerned with interoperability. These public Web sites do not support WMS access to their maps, and they discourage the use of their map tiles in other clients. Choosing a projection which increases interoperability at the cost of visible distortion at large map scales is not to their benefit.

10.5.3 Other Projections

There are a number of other possible projections that tiles may be stored in other than Geodetic and Mercator. Each have their beneficial properties. However, any other projection will also reduce the interoperability of the system. While the Mercator projection does reduce interoperability, it does have better tile client support because it is used by the major Web mapping sites. A single tiled-mapping system using a different projection will probably not have as much support from common geospatial data clients. A non-Geodetic projection will also be more difficult to work with, a problem mentioned with the Mercator projection.

If these problems are not an issue for a tiled-mapping system, there are a number of possible projection which can be used. Primary focus should be on global projections. Most conic projections are not designed to view the entire world so they are less useful for a tiled-mapping system. An equal-area projection, such as the Peter's projection, will create a global map without the area distortions of the Mercator projection or Geodetic projection [2]. The Winkel Tripel projection is the global projection used by the National Geographic Society for their world maps since 1998 [1]. It is neither equal-area nor conformal. Instead, its aim is to reduce, but not eliminate, all map distortions: area, angles, and distance. The Robinson projection is similar to the Winkel Tripel and aims to reduce all distortion rather than eliminate one in particular [5].

It should be noted that UTM is not suitable for a tiled-mapping system. While UTM is a global CRS, it is not a global projection. The different projections defined by UTM for the different sections of the globe will cause serious difficulties for a tiled-mapping system. Tiles will most likely cross UTM boundaries causing great projection distortions, especially at small map scale. Even at large map scales, tiles in different projections will be adjacent and cause discontinuities between tiles.

Only tiled-mapping systems with imagery covering an area completely within a UTM zone should consider using UTM for imagery.

It is also possible to use multiple projections for different tile map scales. For small map scales, use a projection such as Winkel Tripel that does not cause significant size distortion, and for large map scale, use a conformal projection such as Mercator which maintains angles. While possible, this would destroy any interoperability because the tiled-map coordinate system would change depending on map scale. For situations where interoperability is not an issue, this method is viable though it would create significant implementation difficulties.

10.5.4 Which Projection for a Tiled-Mapping System?

Given the importance of interoperability in a tiled-mapping system, we recommend the Geodetic projection as the native projection for image tiles. Clients such as Google Earth and World Wind require input data in the Geodetic projection. Most WMS clients expect map data to be in the Geodetic projection. Other geospatial data, which may be used as image overlays, is often in the Geodetic projection. The simplicity of the Geodetic projection provides additional impetus to prefer it over other projections. The latitude/longitude Cartesian coordinate system makes image manipulation simple, which is important for implementing a WMS front end to the system.

While we recommend the Geodetic projection as the native projection for tiles, it is still possible to provide support for Mercator tiles. The transformation between the Geodetic projection and the Mercator projection is much less complex than Geodetic/UTM reprojections. Rather than store the tiles natively in the Mercator projection, the Mercator tiles may be created as they are requested by the map client. On-the-fly reprojection of Mercator tiles is fast and accurate when using the table-based reprojection method discussed above. An efficient tile cache for the Mercator tiles will ensure that the most commonly requested tiles are available with no additional delay over native Geodetic tiles. If the distortion of the Geodetic projection is a problem, and the primary clients are tile-based, then server-side Mercator reprojection is useful.

On the other hand, if distortion is a problem, but the map client uses continuous scales or a WMS interface, client-side reprojection may be a better option. A client that performs its own reprojection will be able to use any dataset in the Geodetic projection, preserving interoperability. After all the data is retrieved and combined, the new aggregate image may be reprojected once.

If the client performs reprojection, it may choose the target projection best suited for its users. For example, the appropriate UTM projection may be used when the map scale warrants it. Having a client project back to UTM when necessary removes problems of tiles not aligning if stored natively in UTM. Of course, the UTM reprojection is computationally costly and may overwhelm the client system.

   An alternative solution would be to use a simpler reprojection method, which reduces distortion with a simple linear transformation. This simple reprojection method works by scaling a map image to ensure the ground distance for the width and height of a pixel is equal. The scaled image is both equal-area and conformal, though neither is accurate enough for cartographic applications. Code for performing the bounding box modification is shown in Listing 10.1.

   This simple reprojection method can be encoded in the request where the client modifies the geographic bounds to match the latitude and longitude range given by the above formulae. The benefit of encoding the scaling in the request is that it offloads the image manipulation to the server. If the client displays maps using tiles natively, rather than a single composite image created by the server, using this reprojection method becomes more complicated. Using the above formulae on each image tile individually will cause tile alignment issues, much like with the affine transform reprojection. Instead, the formulae should be calculated for a single average latitude for all tiles. Using the same central latitude will ensure the same scaling is performed on each tile. However, further complications arrise as the map view moves north or south and the average latitude for the screen changes. Refreshes of all tiles on the screen will need to occur on a regular basis to ensure the scaling performs adequately.

10.6 Conclusion

Dealing with map projections, datums, and coordinate systems is complicated. This chapter has tried to explain those portions of these topics necessary to properly build a tiled-map system. Our primary focus has been on techniques that provide proper balance to the tiled-mapping system. Tabular point-wise reprojection provides a good balance between accuracy and computations cost. The Geodetic projection provides simplicity and interoperability while allowing client-side adjustments to reduce its distortion.

   Interoperability is a primary concern for most tiled-mapping systems. In general, they are not going to be large systems like Google Maps, Yahoo Maps, or Bing Maps which have no interest in providing data outside of their clients. These systems have the core mission to provide street level mapping and, therefore, an impetus to use the Mercator projection. For tile-mapping systems designed to be used in generic clients using Web services, using a non-Geodetic projection would be a significant impediment to general acceptance. Additionally, with the tabular point-wise reprojection algorithm discussed in this chapter, on-the-fly reprojection from Geodetic tiles to Mercator tiles is certainly feasible. There really is no best projection for all purposes, but the simplicity of the Geodetic projection for computer-based mapping makes it the best choice from an interoperability perspective.

Listing 10.1 Modify a bounding box so that the ground distance for the width and height of a pixel is equal.

```
1  // this method adjust the input bounding box so it will have the same aspect
       ratio as width to height of map display
2  // this is done by adding (never subtracting) area to the bounding box
3  public static BoundingBox adjustBoundingBoxToDimension(BoundingBox bb, int
       width, int height) {
4      BoundingBox adjustedBoundingBox = new BoundingBox(0.0, 0.0, 0.0, 0.0);
5
6      // determine how many longitudinal degrees must be covered for the given
           latitudinal range
7      double averageLatitude = (((bb.maxY + bb.minY) / 2.0) / 360.0) * (2 * Math.
           PI);
8
9      //WGS84
10     double kmInALatitudeDegree = Math.abs(111.13292 − 0.55982 * Math.cos(2.0 *
           averageLatitude) + 0.001175 * Math.cos(4.0 * averageLatitude) −
           0.0000023 * Math.cos(6.0 * averageLatitude));
11
12     double kmInALongitudeDegree = Math.abs(111.41284 * Math.cos(averageLatitude
           ) − 0.0935 * Math.cos(3.0 * averageLatitude) + 0.000118 * Math.cos(5.0
           * averageLatitude));
13
14     // how many degrees of longitude must be covered so that an equal distance
           is covered as is covered by the latitude range
15     double mapAspectRatio = ((double)(width)) / ((double)(height));
16
17
18     double kmCoveredInLatitude = Math.abs(kmInALatitudeDegree * (bb.maxY − bb.
           minY));
19     double kmCoveredInLongitude = Math.abs(kmInALongitudeDegree * (bb.maxX − bb
           .minX));
20
21     double groundDistanceAspectRatio = kmCoveredInLongitude /
           kmCoveredInLatitude;
22
23     if (groundDistanceAspectRatio > mapAspectRatio) { // latitude range must be
           expanded
24         double adjustedLatitudeRange = ((1 / mapAspectRatio) *
               kmCoveredInLongitude) / kmInALatitudeDegree;
25
26         adjustedBoundingBox.minX = bb.minX;
27         adjustedBoundingBox.maxX = bb.maxX;
28         adjustedBoundingBox.minY = (float)(bb.minY − Math.abs((
               adjustedLatitudeRange − Math.abs(bb.maxY − bb.minY)) / 2.0));
29         adjustedBoundingBox.maxY = (float)(bb.maxY + Math.abs((
               adjustedLatitudeRange − Math.abs(bb.maxY − bb.minY)) / 2.0));
30     }
31     else { // longitude range must be expanded
32         double adjustedLongitudeRange = (mapAspectRatio * kmCoveredInLatitude)
               / kmInALongitudeDegree;
33         adjustedBoundingBox.minY = bb.minY;
34         adjustedBoundingBox.maxY = bb.maxY;
35         adjustedBoundingBox.minX = (float)(bb.minX − Math.abs((
               adjustedLongitudeRange − Math.abs(bb.maxX − bb.minX)) / 2.0));
36         adjustedBoundingBox.maxX = (float)(bb.maxX + Math.abs((
               adjustedLongitudeRange − Math.abs(bb.maxX − bb.minX)) / 2.0));
37     }
38
39     return adjustedBoundingBox;
40 }
```

Listing 10.2 Python code to convert a geodetic point to UTM.

```
1  from math import *
2
3  # longitude and latitude should be in degrees
4  # we will convert to radians in the code
5  # longitude should be between [−180, 180)
6  # latitude should be between [−80, 84)
7  def GeodeticToUTM(lon, lat):
8
9      # a and b are WGS84 ellipsoid constants
10     a = 6378137.0
11     b = 6356752.3
12
13     # k0 is a scaling factor used to reduce average distortion
14     k0 = 0.9996
15
16     # the next section of statements calculates the zone
17     # there are some special exceptions
18     if (lat >= 56 and lat < 64 and lon >= 3 and lon < 12):
19         zone = 32
20
21     # this is special zones for Svalbard
22     elif (lat >= 72 and lat < 84):
23         if (lon >= 0 and lon < 9):
24             zone = 31
25         elif (lon >= 9 and lon < 21):
26             zone = 33
27         elif (lon >= 21 and lon < 33):
28             zone = 35
29         elif (lon >= 33 and lon < 42):
30             zone = 37
31
32     # default formula for calculating zone
33     else:
34         zone = int((lon + 180) / 6) + 1
35
36     lonRad = radians(lon)
37     latRad = radians(lat)
38
39     # calculate the central longitude and convert to radians
40     centralLon = (zone − 1) * 6 − 180 + 3
41     centralLonRad = radians(centralLon)
42
43     # do calculations to convert to UTM
44     e = sqrt(1− (b*b)/(a*a))
45     e2 = e*e
46     ePrime2 = (e2) / (1− (e2))
47
48     N = a / sqrt(1 − e2 * pow(sin(latRad), 2))
49     T = pow(tan(latRad), 2)
50     C = ePrime2 * pow(cos(latRad), 2)
51     A = cos(latRad) * (lonRad − centralLonRad)
52
53     M = a * (( 1−e2 / 4 − 3 * pow(e2,2) / 64 − 5 * pow(e2, 3) / 256) * latRad −
                (3 * e2 / 8 + 3 * pow(e2, 2) / 32 + 45 * pow(e2, 3) / 1024) * sin(2*
                latRad) + (15 * pow(e2, 2) / 256 + 45 * pow(e2, 3)/1024) * sin(4*
                latRad) − (35*pow(e2, 3) / 3072) * sin(6*latRad))
54
55     easting = k0 * N * (A + (1−T+C)*pow(A,3)/6+(5−18*pow(T,3)+72*C−58*ePrime2)
                * pow(A,5)/120) + 500000.0
56     northing = k0 * (M +N * tan(latRad) * (pow(A,2) / 2+ (5−T+9*C+4*pow(C,2)) *
                pow(A,4) / 24 + (61 − 58 * T + pow(T,2) + 600 * C − 330 * ePrime2) *
                pow(A,6)/720))
57
58     hemisphere = 'N'
59     # if southern hemisphere then add the false northing
60     if (lat < 0):
```

```
61        northing = northing + 10000000
62        hemisphere = 'S'
63
64
65    return (easting, northing, zone, hemisphere)
66
67
68  for i in xrange(100000000):
69      GeodeticToUTM(-90, 30)
```

References

1. Goldberg, D., Gott, J.: Flexion and skewness in map projections of the Earth. Cartographica: The International Journal for Geographic Information and Geovisualization **42**(4), 297–318 (2007)
2. Iliffe, J.: Datums and Map Projections for remote sensing, GIS, and surveying. Whittles Publishing (2000)
3. Jain, S., Barclay, T.: Adding the EPSG: 4326 Geographic Longitude-Latitude Projection to TerraServer (2003)
4. Mesick, H., Ioup, E., Sample, J.: A Faster Technique for the Transformation of Universal Transverse Mercator Projected. Tech. rep., Naval Research Laboratory (2004)
5. Robinson, A.: A new map projection: Its development and characteristics. International Yearbook of Cartography **14**(1974), 145–155 (1974)
6. Robinson, A., Morrison, J., Muehrcke, P., Kimerling, A., Guptill, S.: Elements of Cartography, 6th edn. John Willey & Sons (1995)
7. Snyder, J.: Map projections: a working manual. USGPO (1987)

Chapter 11
Tile Creation using Vector Data

In previous chapters we were primarily concerned with taking an imagery dataset and converting it into tiles. In this chapter we will focus on vector data as the source for creating tiles. Vector data is made up of geometric primitives, such as points, lines, and polygons. As a result, the process of taking vectors and turning them into image tiles brings up a completely different set of issues from the process using an imagery source.

11.1 Vector Data

As stated above, vector data is geospatial data defined by geometric primitives. A vector dataset will be made up of a number of individual vector features. Each vector feature will have a geometry which defines its geometric shape as well as its location geographically. Often, a vector feature will have a simple geometry made up of a single point, polyline, or polygon (see Figure 11.1). More complex geometries are possible as well. A vector feature may be defined by a curve, a conic, having inner rings, or being multi-dimensional. Additionaly, a vector feature may have a complex geometry that is a combination of other geometries. Examples of complex vector features are shown in Figure 11.2.

In general, there is no guarantee that a vector feature may be drawn into an image or on a computer screen. A curve or a circle must be approximated by a polyline or polygon when drawn by a computer. High-dimensional vector features may not even be approximated for rendering by a computer. Of course, in order to be useful as a source for tiled images, a vector feature must be renderable. As a result we will limit our discussion to features which may be rendered into an image. This chapter will not discuss the process of rendering vector data into an image, which is well described in other texts. Instead, we will limit discussion to the effect the tiling process has on the act of rendering vector features, and vice versa.

J.T. Sample and E. Ioup, *Tile-Based Geospatial Information Systems:*
Principles and Practices, DOI 10.1007/978-1-4419-7631-4_11,
© Springer Science+Business Media, LLC 2010

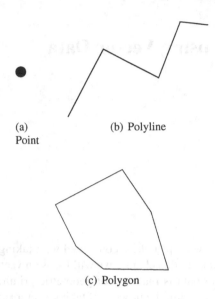

(a)   (b) Polyline
Point

(c) Polygon

Fig. 11.1 Simple vector features.

11.2 Tile Creation

The overall process for creating tiles from vector source data is not significantly different from the process using imagery source data. The primary difference is that instead of cutting image tiles from source data, the vector data is drawn into tiled images. The tile creation process may be outlined in a few simple steps.

1. Choose a tile to create.
2. Determine the bounding box of the tile.
3. Query the vector data source for all features in the bounding box.
4. Render the vector data to an image with the appropriate tile size.
5. Store/disseminate the tile image.

Most of these steps have already been discussed in reference to tiles created from imagery. However, tiling from vector data does have some unique elements which distinguish it from tiling imagery. To recap, the three primary differences are:

- Storage space: Rendered image tiles require a significant amount of storage space relative to vector map content. A collection of geospatial features might be 100 megabytes in vector form but could grow to several terabytes when rendered over several different scales.
- Processing time: Pre-rendering image tiles requires a significant amount of time, and many of those tiles may be in geographic areas of little interest to users. The

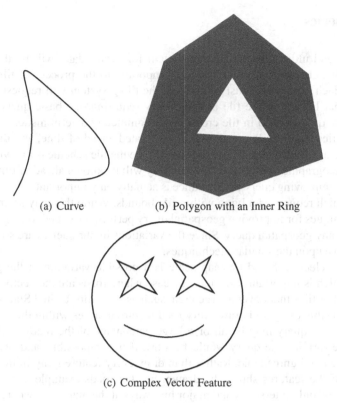

(a) Curve   (b) Polygon with an Inner Ring

(c) Complex Vector Feature

Fig. 11.2 More complicated vector features.

most efficient way to decide what tiles to render is to wait until they are requested by actual users.

- Overview images: Overview images, i.e., very low scale images, can be rendered directly from geospatial vectors. Unlike raster-based tile systems, there is no need to render the high scale views first and then generate scaled down versions.

These differences allow a modification of how the resulting tiles created from vectors are used. For imagery, all the tiles are created ahead of time and stored permanently for distribution. However, because of the above three differences in tiled imagery, it does not make sense to pre-render all tiles ahead of time. Instead, the tiles may be rendered as they are requested by users. By rendering tiles just-in-time, only the tiles which users request are ever generated. On-the-fly rendering reduces storage requirements significantly but will reduce performance of the system. Speed is improved if the rendered tiles are cached on disk after they are first requested. Each tile is only rendered once (or until the underlying data changes). The most commonly viewed tiles will be cached so performance will be good.

11.3 Queries

Querying, and ultimately the storage system for vector data, will be the primary focus of this chapter. Only one query is important to the process of tiling vector features. Each time a tile must be created, the tiling system will request all vector features that lie within the tile bounds. Only requiring one basic query is useful because the queries used in tile creation are completely deterministic. A complete list of queries used in tile creation may be created ahead of time, and the list will not change over repeated tile creation runs. Since the tile scheme is known ahead of time, the geographic bounds used in the query will be known ahead of time as well. Therefore, improving query performance is actually only important for the subset of queries which request data lying within tile bounds. While this may seem obvious, most techniques for improving geospatial query performance tend to be generalized to support any geospatial query. Since the variations in our queries are so small, we can improve upon the standard techniques.

Feature selection based on map scale is a possible variation to the geospatial queries which is important. Consider the case where roads are the vector data used for tiling. For tiles that cover a large area, such as the entire United States, it would be unreasonable to try and draw every road feature that lies within the tile. The results of such a query may be all or a large percentage of the roads in the overall dataset. Performing the query would be slow, if not impossible, and the rendered tile cluttered and unreadable. Rather than draw every feature lying in the tile, only a subset of the features should be drawn. For the roads example, it would make sense to draw only interstates and major highways at the map scale where the entire United States is visible. In general, in cases where the vector data source is quite large, consideration should be given to selecting features to draw based on the map scale of the destination tile. For roads, we may choose to have interstates and major highways drawn at all zoom levels, minor highways drawn at zoom level 5 and above, and all roads drawn at zoom level 8 and above. Using map scale, or zoom level, to select features does add complexity the queries used to generate tiles. However, it will generally improve performance because lower zoom level tiles will have fewer features to retrieve. Filtering features based on scale also does not change the fact that the tiling queries are completely deterministic. The determinism makes it possible to design a highly targeted storage methodology for tiling vector data.

11.4 Storage

There are two primary methods of storing vector data for tiling: database storage and file system storage. Database storage of vector data is more common than file storage when the data is to be retrieved using geospatial queries. File storage is more commonly used for archival and distribution of vector data as fixed data sets. We will describe a file storage system which is designed to support high performance tile creation from vector data.

11.4.1 Database Storage

Most modern database systems provide support for geospatial data, including storing geometries as first class data types, supporting complex geospatial queries, and providing geospatial indexes. Common examples include Oracle with the Spatial extension, PostgreSQL/PostGIS, and MySQL. Normally, vector data is stored in a database with each attribute of the feature appearing as a field in a database table, including the geometry. Because database tables have a fixed schema, features must have the same attributes to be stored in the same table. As a result, there is usually a direct mapping between geospatial layers and database tables. Many tools exist to import data from geospatial file formats directly into tables (such as shp2pgsql for PostgeSQL).

Querying for geospatial data is supported by these databases without any extra development. The query for all features in the geospatial bounds of a tile is shown in Listing 11.1.

Listing 11.1 Geospatial query for vector features within a tile's bounding box. Based on PostgreSQL/PostGIS.

```
1  SELECT * FROM FeaturesTable WHERE feature_geometry && ST_MakeBox2D(ST_Point
       (-90, 0),ST_Point(0, 90));
```

The operator && determines whether the bounding boxes of two geometries overlap. In this example, the query is comparing the feature geometry with the tile (1, 1) at zoom level 2.

Indices should be created to provide adequate performance of these queries. The two most commonly used geospatial indices are the R-Tree index (usually the R*-Tree variant) and the Quadtree. The R-Tree index tends to be preferred over the Quadtree because it provides better query performance over a wider variety of geospatial queries. The database controls the creation of indices, though some level of tuning is allowed by the user. Regardless of what type of geospatial index is used, the database tables should be clustered. Clustered data is ordered on disk according to its location in the index. As a result, the database must access only a localized area of the disk when solving a query using the index. Index clustering is essential to query performance when the query is expected to return multiple results. An example of creating a clustered index in PostgreSQL/PostGIS is shown in Listing 11.2.

Listing 11.2 Example SQL for creating a clustered PostGIS R-Tree index in PostgreSQL.

```
1  CREATE INDEX spatial_index ON FeaturesTable USING GIST ( feature_geometry );
```

Creating the database tables from vector data layers and using clustered indices are sufficient to create a functioning database environment for vector tiling. No changes to the geospatial query are necessary. A database will automatically determine if an index should to be used in the evaluation of a query.

Additional modifications to the geospatial database may be implemented to increase performance. Given the foreknowledge of the query patterns to the database, we can customize the way the features are stored to simplify evaluation for the known queries. The first case to examine is the one where only a subset of features are used in tiles with low zoom level. A simple modification to the query in Listing 11.2 will add support for this functionality, as seen in Listing 11.3.

Listing 11.3 Geospatial query with filter on zoom level.

```
1 SELECT * FROM FeaturesTable WHERE feature_geometry && ST_MakeBox2D(ST_Point
    (-90, 0),ST_Point(0, 90)) AND feature_min_zoom_level <= 2;
```

This query will accomplish the goal of retrieving only a subset of the features inside a tile, depending on the scale. However, there is a problem. The query on bounds and scale requires different optimizations depending on the query parameters. When the query zoom level is high and the query bounds are small, a clustered geospatial index will provide the best performance. On the other hand, when the query zoom level is small and the query bounds are large, a clustered zoom level index will provide the best performance. Given that only one index on a table may be clustered, it is inevitable that one of these two sets of queries will not perform as efficiently as possible.

The solution is simple: create multiple tables to hold data for different scales. The key zoom levels where the subset of tiles being rendered changes are known ahead of time. A separate table may be created to hold the features whose minimum display zoom level is less than a particular key zoom levels. For example, if the key zoom levels are zero, eight, and fourteen then our roads layer will have tables Roads_0, Roads_8, and Roads_14. Each of these tables holds all features whose minimum zoom level is less than or equal to the table zoom level. The tiler need only query the table which matches the zoom level of the tile being rendered at the moment. No additional filtering on zoom level need be done in the query because the filtering has been performed ahead of time.

The only drawback to creating multiple tables for each layer is that features will be duplicated between tables. For example, interstates should be represented in all three roads tables. However, additional performance is possible by reducing the resolution (i.e. the number of points in the geometry) of features in the tables with lower zoom level. The reduced resolution will decrease the size of the table and increase the speed of rendering a tile. Creating multiple tables increases the required space to hold the vector data, but disk is cheap, and vector data is relatively small, especially in comparison to imagery. The performance benefits of creating a table for each key zoom level far outweigh the storage costs. Figure 11.3 shows different key zoom levels used by OpenStreetMap.

One of the benefits of a database is that it automatically organizes and searches a wide variety of data types. A developer can store data in a database with little or no custom development. The flip side of the automatic and general nature of a database is the limited amount of customization that is possible. The R-Tree index included in a database is a good example of this tradeoff. The R-Tree index is a good geospatial index which increases performance of a wide variety of geospatial queries. However, the database completely manages the organization of data in the R-Tree index. The application developer has no way of guiding the organization of data in the R-Tree. We know ahead of time which vector features lie within which

Fig. 11.3 Key zoom levels used by OpenStreetMap when rendering their vector data.

tiles. It would provide a performance benefit if the application developer could ensure that the R-Tree page splits matched tile splits, but this is not possible. Such customization would be difficult even in a Quadtree, which organizes data into tiles by default.

The loss of customization inherent in using a database would be acceptable if the database provided significant advantages for a vector tiling system. However, a tiling system does not require much of the functionality provided by a database. Core database features like advanced locking of data and rollback availability are unnecessary for a vector datastore which is primarily read-only. These database features are not free; they are a core part of the database which are included at a cost to performance. For example, a banking system which requires accuracy in monetary transactions definitely requires atomic transactions in its datastore.

11.4.2 File System Storage

In contrast to a database, file system storage offers little in the way of automatic functionality but provides the developer with the ability to fully customize the storage implementation in the overall system. A tiling system is a good example of an application which can benefit from using a file system for storage. The deterministic nature of the queries performed when tiling features provides an environment which can benefit from the customization allowed by a file system. Using our knowledge of the vector tiling system, we can develop a custom file storage implementation which is optimized for our system.

The first departure in our file storage implementation from the database design is in how vector layers are managed. In the database, each layer is mapped to a separate table. This design is necessary because database tables have a fixed schema which requires all records to share the same columns. In general, different vector layers will not have the same feature attributes (which map to table columns) and, as a result, must be stored in different tables. A custom file format does not have this restriction. The file store may be designed to hold features from many different layers. There is a good reason to store features from multiple layers in the same file. The tile system uses features from multiple layers when drawing tiles. The design of a custom storage system should partition data only when it benefits the overall efficiency of the system. Usually, the performance benefit comes from the query access patterns. Since the queries in the tiling system do not require features partitioned by layers, there is no reason to do so. The simplest way to store features from multiple layers in the same file is to store a variable size list of the attribute names and values for each feature. This method obviously requires more storage space than fixed schema systems. As a means to reduce storage requirements, the attribute names for the different feature layers may be stored once in the file header and linked to each feature. However, storage is generally cheap so the added complexity may not be worth the effort.

In contrast, the geospatial area is a property of the vector features that affects query access patterns. Therefore, partitioning data according to location is important for an efficient file storage system. Databases improve the performance of the geospatial queries by using a geospatial index. We already mentioned that by automatically creating these indices, the database would never provide a geospatial index optimized for our tiling application. With a custom file storage implementation, we can optimize storage for our tiling application. We can take advantage of the fact that geospatial queries used by the tiling system always match tile boundaries. As a result, we index and cluster the vector data using tile boundaries.

The simplest index partitions the data into multiple files whose bounds align with the tile bounds at one chosen zoom level. However, as we have seen with image tiles, at high zoom levels the number of files becomes unwieldy. Instead, we partition the data according to tile location but store each partition in one file. The start byte and length of every tile partition is stored separately so that each tile may be accessed independently. A feature is placed into a tile partition if its geospatial bounds overlap the bounds of the file's corresponding tile (see Figure 11.4). It is likely that a few of the features will overlap multiple tiles. In this case, the features are placed into each overlapping tile partition. The result is a file storage scheme which is by default also a clustered index. Using features stored on the file system is easy. An entire tile partition may be loaded into memory, and the tile it represents is rendered. All subtiles at higher zoom levels may be rendered as well. Alternatively, features may be rendered as they are read from the file system without caching them in memory, allowing lower resource systems to use the same scheme.

To support memory caching, the tile partitions must be sized to fit into memory. Thus, the zoom level which defines the boundaries of the tile partitions should be the lowest zoom level whose tile partitions fit into the memory of the rendering system. Determining the appropriate target zoom level will require some experimentation, but if performance is a concern, the results will be worth it.

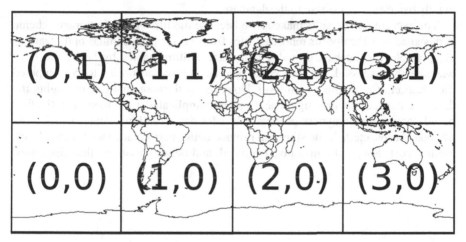

Fig. 11.4 A map made up of polyline vector features. Each polyline is partitioned according to which zoom level 2 tile it lies within. For example, Antarctica would be placed in tiles (0,0), (1,0), (2,0), and (3,0).

Minimum rendering zoom level is another vector feature property which may be managed by partitioning data. For database storage, a table was made for each key zoom level in the tiling system. The same technique may be used for file storage. Each zoom level which uses a different subset of features for rendering has a separate file to store features. That directory stores the feature file and its index. The files for a key zoom level are used when rendering tiles at that zoom level or higher (until the next higher key zoom level).

As with the database version of this optimization, overall storage cost is increased by redundantly storing features. Conversely, the average amount of data accessed when performing a query is reduced because there are fewer features in each file. The result of this custom vector data store is that all queries are essentially precomputed so that the disk accesses are all predetermined. Each file will only be read off disk once, and all tiles may be rendered by looking at only one file. Once data is in memory, the cost of filtering features to render tiles with smaller geographic areas is small. Disk access is much more costly than in-memory computation.

Experimentation shows that the performance of a file-based feature store out performs a standard database. The feature data used for the testing is the road network of the United States. The data comes from the NAVTEQ corporation and is the same dataset used by the commercial Web-mapping systems. We created a basic file-based storage system with features partitioned at zoom level 11 and stored in a file. We also created a PostgreSQL/PostGIS datasbase to store the same features. The database table was clustered using an R-Tree index on the data. The experiment query requests all the features in a tile. The tests were performed using a random list of tiles from zoom level 11 located in the continental United States. The time to execute each query and the number of features in the queried tile was recorded. The results, as seen in Figure 11.5, show that the queries to the file store are approximately twice as fast as those to the database.

The experimental results make sense because the file-based tile storage scheme is designed specifically for rendering tiles from vector data. Similar optimizations are possible for database stored vector data; splitting tables by key zoom level was already discussed, but data could also be indexed according to precomputed tile location at a specific zoom level. However, with these changes, managing the database storage becomes significantly more complicated, even more than the file-based storage (querying multiple layer tables for data, handling features which cross tile boundaries, etc.). A file store can provide better performance with lower development cost, lower administrative overhead, and better portability than databases.

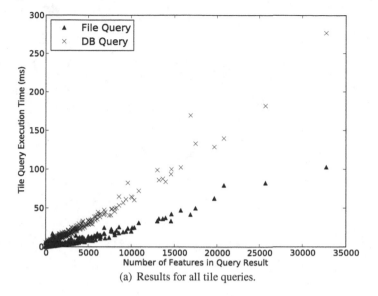

(a) Results for all tile queries.

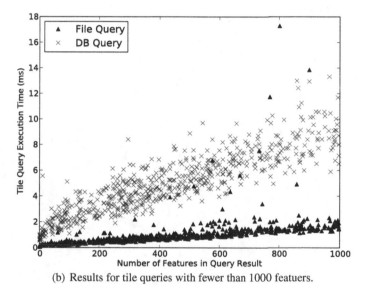

(b) Results for tile queries with fewer than 1000 featuers.

Fig. 11.5 Comparison of geospatial query execution times between a database and a file-based feature store. Each geospatial query requests all features in a tile from zoom level 11. The file query outperforms the database, most significantly when the number of features in the query tile grows.

Chapter 12
Case Study: Tiles from Blue Marble Imagery

In this chapter we will present a complete end-to-end system for creating and storing tiled images from a freely available worldwide set of imagery. The system will read source imagery, cut it into tiled images, and store the tiled images to cluster files.

NASA's Blue Marble Next Generation Imagery (BMNG) is a composite image of the Earth at 500 meters resolution taken by the MODIS satellite mounted sensor. The BMNG imagery and information about it are freely available for download from

http://earthobservatory.nasa.gov/Features/BlueMarble/

The imagery comes in two formats: as a single raw image file 86,400 pixels wide by 43,200 pixels high and as 8 smaller sub-images, 21,600 pixels by 21,600 pixels. In this chapter we will present a pull-based tiling approach using the single large image and a push-based tiling approach using the 8 sub-images.

Before we can begin tiling, we must determine the base zoom level that we will use for our tile set. Both the single large image and the set of 8 sub-images have the same geospatial and image resolution, so we will use the same base zoom level for both image sets. Using the following equation, we can compute the degrees per pixel for our Blue Marble imagery.

$$(360.0/84600 + 180.0/42300)/2 = 0.00425.$$

Since 0.00425 falls between level 7 (0.00549) and level 8 (0.00274), as shown in Table 2.1, we will choose level 8 as our base level.

12.1 Pull-Based Tiling

The algorithm presented in this section will bring together six concepts already presented in the book:

J.T. Sample and E. Ioup, *Tile-Based Geospatial Information Systems:* *Principles and Practices*, DOI 10.1007/978-1-4419-7631-4_12,
© Springer Science+Business Media, LLC 2010

- Section 5.2.3: Pull-based tile creation that iterates over the tiles first. For each tile, it extracts the required data from the source images, creates the tile, and stores the tile.
- Section 5.3.1: Scaling process for lower resolution tiles.
- Section 6.1: An optimized version of tile creation that holds tiles in memory while they are being used and write them to disk in memory.
- Section 6.2.1: Methods for partial reading of source images.
- Section 6.3.1: Multi-threading tile creation.
- Section 8.5: Storage of tiled images in clusters of tiles from different zoom levels.

Because our source image is too large to hold in memory all at once, we will implement an algorithm for partial reading of the image. The image, like many images, is stored in row-major order, also known as scanline order. The Java class in Listing 12.1 provides a method for reading pixel delineated sub-sections of the large Blue Marble image. The example code is designed for clarity and ease of understanding. There are more efficient ways to read sub-images. These include setting pixels in blocks of data versus setting one pixel at a time and reading blocks of bytes instead of one byte at a time. For simplicity's sake, we will multi-thread the creation of only the base level. The higher levels take a much shorter amount of time to create and do not require multi-threading.

The next piece of supporting code we will need is a modified version of the tile cluster storage algorithm. The version in Listing 12.2 takes the code presented in Section 8.5 and adds in-memory caching of tiled images during the creation process and a cache of open RandomAccessFiles. Since this section is primarily concerned with creating tiles, the code only manages a write cache. Reading of tiles stored to disk in an earlier session is always done directly from disk. Reading of tiles that have just been written to the cache is done from the cache. Also, the write cache uses a hashmap with Java String objects as keys. A more efficient approach would use numerical tile addresses as keys, but the String based approach is simpler to implement. The final piece of supporting code, Listing 12.3, allows multiple threads to iterate over a range of tile addresses.

Given the supporting code, now we can create the completed system. The steps in the algorithm are as follows:

1. Iterate over all the tiles in the base zoom level. For each tile:

   a. Pull the imagery needed from the source image.
   b. Scale the tile to the proper resolution.
   c. Store the tile in the cache.

2. Iterate over each successive zoom level up to level 1.

   a. For each zoom level, iterate over each tile at that level. For each tile:
      i. Pull the four images from the higher level that make up the current tile.
      ii. Merge the four images together into one image.
      iii. Store that image into cache.

3. After all tiles have been created, write the tiles to disk.

The Java classes in Listing 12.4 demonstrate this algorithm. The first class, PullTile-Creation, initializes the input and output, creates and starts the pull tiler threads, and creates the lower resolution levels. The class PullTilerThread is a Java thread implementation that does the work of creating the base level tiled images.

12.2 Push-Based Tiling

In this example, we will use as source imagery the Blue Marble data that has been divided into 8 sub-images. Each image is 21,600 by 21,600 pixels and covers a 90 degree by 90 degree area of the earth's surface. Since each sub-image can be held completely in memory, we can use a push-based tiling approach. Our algorithm will iterate over the source images in memory and extract data from each source image needed to make the tiled images.

The algorithm presented in this section will bring together four concepts already presented in the book:

- Section 5.2.3: Push-based tile creation that iterates over the source images first.
- Section 5.3.1: Scaling process for lower resolution tiles.
- Section 6.1: An optimized version of tile creation that held tiles in memory while they were being used and wrote them to disk in memory.
- Section 8.5: Storage of tiled images in clusters of tiles from different zoom levels.

It should be noted that the two techniques differ only in the method of creating the base level tiles. The section of code for creating the lower resolution levels is exactly the same as in the previous section. Listing 12.5 shows the push-based method for creating the Blue Marble tiles.

12.3 Results

Each technique creates the required tile clusters, and both methods gave practically identical results. The multi-threaded pull-based method took 1,284.05 seconds, while the single threaded push-based method took 1,553.90 seconds. The top two tiles from the completed sets are presented in Figures 12.1 and 12.2.

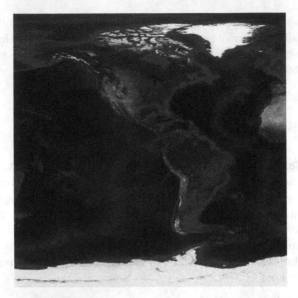

Fig. 12.1 Tile (0,0) from Blue Marble.

Fig. 12.2 Tile (0,1) from Blue Marble.

Listing 12.1 Example code for reading raw Blue Marble imagery.

```java
public class RawImageReader {

    private String filename;
    int imageWidth;
    int imageHeight;
    private RandomAccessFile raf;
    private long bytesPerRow;
    public BoundingBox imageBounds;

    public RawImageReader(String filename, int width, int height, BoundingBox
            imageBounds) {
        this.filename = filename;
        try {
            this.raf = new RandomAccessFile(filename, "r");
        } catch (FileNotFoundException e) {
            e.printStackTrace();
        }
        this.imageWidth = width;
        this.imageHeight = height;
        this.imageBounds = imageBounds;
        this.bytesPerRow = width * 3;
    }

    public synchronized BufferedImage getSubImage(int startx, int starty, int
            width, int height) {
        try {
            // create an empty image
            BufferedImage bi = new BufferedImage(width, height, BufferedImage.
                TYPE_INT_RGB);
            // image is stored in row-major order
            for (int j = 0; j < height; j++) {
                // determine start position of the row to be read
                long startPosition = (starty + j) * bytesPerRow + startx * 3;
                // seek to the portion of the row that we need
                this.raf.seek(startPosition);

                for (int i = 0; i < width; i++) {
                    int r = this.raf.read();
                    int g = this.raf.read();
                    int b = this.raf.read();
                    // combine the rgb byte values into a single int value
                    int rgb = 0xff000000 | r << 16 | g << 8 | b;
                    // set the image pixel to the combined rgb value
                    bi.setRGB(i, j, rgb);
                }
            }
            return bi;
        } catch (IOException e) {
            e.printStackTrace();
        }
        return null;

    }


    public void close() {
        try {
            this.raf.close();
        } catch (IOException e) {
            e.printStackTrace();
        }
    }
```

Listing 12.2 Cached clustered tile I/O.

```
1  public class CachedClusteredTileStream {
2
3      static final long magicNumber = 0x772211ee;
4      private String location;
5      private String setname;
6      private int numlevels;
7      private int breakpoint;
8      HashMap < String ,
9      BufferedImage > writeCache = new HashMap < String ,
10     BufferedImage > ();
11     HashMap < String ,
12     RandomAccessFile > openFileCache = new HashMap < String ,
13     RandomAccessFile > ();
14
15     public CachedClusteredTileStream (String location , String setname , int
            numlevels , int breakpoint) {
16         this.location = location;
17         this.setname = setname;
18         this.numlevels = numlevels;
19         this.breakpoint = breakpoint;
20     }
21
22     public void writeTile (long row, long column, int level , BufferedImage image
            ) {
23         String key = row + ":" + column + ":" + level;
24         writeCache.put(key, image);
25     }
26
27     public BufferedImage readTile (long row, long column, int level) {
28         String key = row + ":" + column + ":" + level;
29         if (writeCache.containsKey(key)) {
30             return writeCache.get(key);
31         } else {
32             ByteArrayInputStream bais = new ByteArrayInputStream(
33             readTileFromDisk(row, column, level));
34             BufferedImage bi = null;
35             try {
36                 bi = ImageIO.read(bais);
37             } catch (IOException e) {
38                 e.printStackTrace ();
39             }
40
41             return bi;
42         }
43     }
44
45     public void writeTileFromCache(long row, long column, int level , byte[]
            imagedata) {
46
47         //first determine the cluster that will hold the data
48         ClusterAddress ca = getClusterForTileAddress (row, column, level);
49         String clusterFile = getClusterFileForAddress (ca);
50         if (clusterFile == null) {
51             return;
52         }
53         File f = new File(clusterFile);
54
55         //if the file doesn't exist , set up an empty cluster file
56         if (!f.exists()) {
57             createNewClusterFile(f, ca.endLevel - ca.startLevel + 1);
58         }
59         try {
60
61             RandomAccessFile raf = getOpenFileFromCache(f);
62
63             //write the tile and info at the end of the tile file
```

```
64          long tilePosition = raf.length();
65          raf.seek(tilePosition);
66          raf.writeLong(magicNumber);
67          raf.writeLong(magicNumber);
68          raf.writeLong(column);
69          raf.writeLong(row);
70          raf.writeInt(imagedata.length);
71          raf.write(imagedata);
72
73          // determine the position in the index of the tile address
74          long indexPosition = getIndexPosition(row, column, level);
75          raf.seek(indexPosition);
76
77          // write the tile position and size in the index
78          raf.writeLong(tilePosition);
79          raf.writeInt(imagedata.length);
80
81      } catch (Exception e) {
82          e.printStackTrace();
83      }
84  }
85
86  public byte[] readTileFromDisk(long row, long column, int level) {
87      // first determine the cluster that will hold the data
88      ClusterAddress ca = getClusterForTileAddress(row, column, level);
89      String clusterFile = getClusterFileForAddress(ca);
90      if (clusterFile == null) {
91          return null;
92      }
93      File f = new File(clusterFile);
94
95      try {
96          RandomAccessFile raf = getOpenFileFromCache(f);
97
98          // determine the position in the index of the tile address
99          long indexPosition = getIndexPosition(row, column, level);
100         raf.seek(indexPosition);
101         long tilePosition = raf.readLong();
102         int tileSize = raf.readInt();
103         if (tilePosition == -1L) {
104             // tile is not in the cluster
105             raf.close();
106             return null;
107         }
108         byte[] imageData = new byte[tileSize];
109         // offset tile position for header information
110         long tilePositionOffset = tilePosition + 8 + 8 + 8 + 8 + 4;
111         raf.seek(tilePositionOffset);
112         raf.readFully(imageData);
113
114         return imageData;
115     } catch (Exception e) {
116         e.printStackTrace();
117     }
118     return null;
119 }
120
121 private long getIndexPosition(long row, long column, int level) {
122     ClusterAddress ca = this.getClusterForTileAddress(row, column, level);
123     // compute the local address, that's the relative address of the tile in
              the cluster
124     int localLevel = level - ca.startLevel;
125     long localRow = (long)(row - (Math.pow(2, localLevel) * ca.row));
126     long localColumn = (long)(column - (Math.pow(2, localLevel) * ca.column
              ));
127     int numColumnsAtLocalLevel = (int) Math.pow(2, localLevel);
```

```
128        long indexPosition = this.getCumulativeNumTiles(localLevel − 1) +
                 localRow * numColumnsAtLocalLevel + localColumn;
129        // multiply index position times byte size of a tile address
130        indexPosition = indexPosition * (8 + 4);
131        return indexPosition;
132    }
133
134    public ClusterAddress getClusterForTileAddress(long row, long column, int
           level) {
135        if (level > this.numlevels) {
136            // error, level is outside of ok range
137            return null;
138        }
139        int targetLevel = 0;
140        int endLevel = 0;
141        if (level < breakpoint) {
142            // tile goes in one of top two clusters
143            targetLevel = 1;
144            endLevel = breakpoint − 1;
145        } else {
146            // tile goes in bottom cluster
147            targetLevel = this.breakpoint;
148            endLevel = this.numlevels;
149        }
150        // compute the difference between the target cluster level and the tile
               level
151        int powerDiff = level − targetLevel;
152        // level factor is the number of tiles at level "level" for a cluster
               that starts at "target level"
153        double levelFactor = Math.pow(2, powerDiff);
154        // divide the row and column by the level factor to get the row and
               column address of the cluster we are using
155        long clusterRow = (int) Math.floor(row / levelFactor);
156        long clusterColumn = (int) Math.floor(column / levelFactor);
157        ClusterAddress ca = new ClusterAddress(clusterRow, clusterColumn,
               targetLevel, endLevel);
158        return ca;
159    }
160
161    String getClusterFileForAddress(ClusterAddress ca) {
162        String filename = this.location + "/" + this.setname + "−" + ca.
               startLevel + "−" + ca.row + "−" + ca.column + ".cluster";
163        return filename;
164    }
165
166    // this methods create an empty file and fills the index with null values
167    void createNewClusterFile(File f, int numlevels) {
168        RandomAccessFile raf;
169        try {
170            raf = getOpenFileFromCache(f);
171            raf.seek(0);
172            long tiles = this.getCumulativeNumTiles(numlevels);
173            for (long i = 0; i < tiles; i++) {
174                raf.writeLong(−1L); // NULL position of tile
175                raf.writeLong(−1L); // NULL size of tile
176            }
177        } catch (Exception e) {
178            e.printStackTrace();
179        }
180    }
181
182    public int getCumulativeNumTiles(int finallevel) {
183        int count = 0;
184        for (int i = 1; i <= finallevel; i++) {
185            count += (int)(Math.pow(2, 2 * i − 2));
186        }
187        return count;
```

```
188         }
189
190         public RandomAccessFile getOpenFileFromCache( File f) {
191             String key = f.getAbsolutePath();
192             if (openFileCache.containsKey(key)) {
193                 return openFileCache.get(key);
194             } else {
195                 try {
196                     RandomAccessFile raf = new RandomAccessFile(f, "rw");
197                     openFileCache.put(key, raf);
198                     return raf;
199                 } catch (FileNotFoundException e) {
200                     e.printStackTrace();
201                 }
202             }
203             return null;
204         }
205
206         public void close() {
207             // iterate over tiles in the cache and write them to disk
208             Set < String > keys = writeCache.keySet();
209             for (String s: keys) {
210                 String[] data = s.split(":");
211                 long row = Long.parseLong(data[0]);
212                 long column = Long.parseLong(data[1]);
213                 int level = Integer.parseInt(data[2]);
214                 BufferedImage image = writeCache.get(s);
215                 ByteArrayOutputStream baos = new ByteArrayOutputStream();
216                 try {
217                     ImageIO.write(image, "jpg", baos);
218                 } catch (IOException e) {
219                     e.printStackTrace();
220                 }
221                 byte[] imagedata = baos.toByteArray();
222                 writeTileFromCache(row, column, level, imagedata);
223             }
224             Set < String > openFiles = openFileCache.keySet();
225             for (String f: openFiles) {
226                 RandomAccessFile raf = openFileCache.get(f);
227                 try {
228                     raf.close();
229                 } catch (IOException e) {
230                     e.printStackTrace();
231                 }
232             }
233         }
234 }
```

Listing 12.3 Thread Safe Tile Range Iterator.

```
1  public class TileRangeIterator {
2
3      long curcol,
4      currow,
5      maxrow,
6      maxcol,
7      mincol,
8      minrow;
9      int level;
10
11     public TileRangeIterator(long minrow, long maxrow, long mincol, long maxcol
           , int level) {
12         this.minrow = minrow;
13         this.maxrow = maxrow;
14         this.mincol = mincol;
15         this.maxcol = maxcol;
```

```
16
17          this.curcol = mincol;
18          this.currow = minrow;
19          this.level = level;
20
21      }
22
23      public boolean hasMoreTiles() {
24          if ((this.currow <= this.maxrow)) {
25              return true;
26          }
27          return false;
28      }
29
30      public synchronized TileAddress getNextTileID() {
31          TileAddress address = new TileAddress(this.currow, this.curcol, this.
                level);
32          this.curcol++;
33          if (this.curcol > this.maxcol) {
34              this.currow++;
35              this.curcol = this.mincol;
36          }
37          return address;
38
39      }
40 }
```

Listing 12.4 Pull-based tile creation for Blue Marble.

```
1  public class PullTileCreation {
2
3      static int TILE_SIZE = 512;
4
5      public static void main(String[] args) {
6
7          BoundingBox imageBounds = new BoundingBox(-180, -90, 180, 90);
8          int imageWidth = 86400;
9          int imageHeight = 43200;
10
11         int baselevel = 8;
12         int numthreads = 8;
13         CachedClusteredTileStream cts = new CachedClusteredTileStream("folder",
               "bluemarble", baselevel, baselevel + 1);
14
15         RawImageReader imageReader = new RawImageReader("world.topo.bathy
               .200407.3x86400x43200.bin", imageWidth, imageHeight, imageBounds);
16
17         // initilize values for base level
18
19         long startRow = 0;
20         long startColumn = 0;
21         long endRow = TileStandards.zoomRows[baselevel] - 1;
22         long endColumn = TileStandards.zoomColumns[baselevel] - 1;
23
24         // build tiles for base level
25
26         // initilize the tile range iterator
27         TileRangeIterator tri = new TileRangeIterator(startRow, endRow,
               startColumn, endColumn, baselevel);
28
29         // create and start the tiling threads
30         Thread[] threads = new Thread[numthreads];
31         for (int i = 0; i < threads.length; i++) {
32             threads[i] = new PullTilerThread(tri, cts, imageReader);
33             threads[i].start();
34         }
```

```
35        //wait for the threads to finish
36        for (int i = 0; i < threads.length; i++) {
37            try {
38                threads[i].join();
39            } catch (InterruptedException e) {
40                e.printStackTrace();
41            }
42        }
43
44        //iterate over the remaining levels
45        for (int level = baselevel - 1; level >= 1; level--) {
46            int ratio = (int) Math.pow(2, baselevel - level);
47            long curMinCol = (long) Math.floor(startColumn / ratio);
48            long curMaxCol = (long) Math.floor(endColumn / ratio);
49            long curMinRow = (long) Math.floor(startRow / ratio);
50            long curMaxRow = (long) Math.floor(endRow / ratio);
51            //Iterate over the tile set coordinates.
52            for (long c = curMinCol; c <= curMaxCol; c++) {
53                for (long r = curMinRow; r <= curMaxRow; r++) {
54                                    //For each tile, do the following:
55                    TileAddress address = new TileAddress(r, c, level);
56                    //Determine the FOUR tiles from the higher level that
                         contribute to the current tile.
57                    TileAddress tile00 = new TileAddress(r * 2, c * 2, level +
                         1);
58                    TileAddress tile01 = new TileAddress(r * 2, c * 2 + 1,
                         level + 1);
59                    TileAddress tile10 = new TileAddress(r * 2 + 1, c * 2,
                         level + 1);
60                    TileAddress tile11 = new TileAddress(r * 2 + 1, c * 2 + 1,
61                                                         level + 1);
62                    //Retrieve the four tile images, or as many as exist.
63
64                    BufferedImage image00 = cts.readTile(tile00.row, tile00.
                         column,
65                                                         tile00.level);
66                    BufferedImage image01 = cts.readTile(tile01.row, tile01.
                         column,
67                                                         tile01.level);;
68                    BufferedImage image10 = cts.readTile(tile10.row, tile10.
                         column,
69                                                         tile10.level);
70                    BufferedImage image11 = cts.readTile(tile11.row, tile11.
                         column,
71                                                         tile11.level);
72                    //Combine the four tile images into a single, scaled-down
                         image.
73                    BufferedImage tileImage = new BufferedImage(
74                                                         TILE_SIZE,
75                                                         TILE_SIZE,
76                                                         BufferedImage.
                                                             TYPE_INT_RGB
                                                         );
77                    Graphics2D g = (Graphics2D) tileImage.getGraphics();
78                    g.setRenderingHint(RenderingHints.KEY_INTERPOLATION,
79                                       RenderingHints.
                                           VALUE_INTERPOLATION_BILINEAR);
80                    boolean hadImage = false;
81                    if ((image00 != null)) {
82                        g.drawImage(image00, 0, Constants.TILE_SIZE_HALF,
83                                    Constants.TILE_SIZE_HALF, Constants.
                                        TILE_SIZE,
84                                    0, 0, Constants.TILE_SIZE, Constants.
                                        TILE_SIZE,
85                                    null);
86                        hadImage = true;
87                    }
```

```
 88                      if ((image01 != null)) {
 89                          g.drawImage(image01, Constants.TILE_SIZE_HALF,
 90                                      Constants.TILE_SIZE_HALF, Constants.
                                             TILE_SIZE,
 91                                      Constants.TILE_SIZE, 0, 0, Constants.
                                             TILE_SIZE,
 92                                      Constants.TILE_SIZE, null);
 93                          hadImage = true;
 94                      }
 95                      if ((image10 != null)) {
 96                          g.drawImage(image10, 0, 0, Constants.TILE_SIZE_HALF,
 97                                      Constants.TILE_SIZE_HALF, 0, 0,
 98                                      Constants.TILE_SIZE, Constants.TILE_SIZE,
                                             null);
 99                          hadImage = true;
100                      }
101                      if ((image11 != null)) {
102                          g.drawImage(image11, Constants.TILE_SIZE_HALF, 0,
103                                      Constants.TILE_SIZE, Constants.
                                             TILE_SIZE_HALF,
104                                      0, 0, Constants.TILE_SIZE, Constants.
                                             TILE_SIZE,
105                                      null);
106                          hadImage = true;
107                      }
108                      //save the completed tiled image to the tile storage
                             mechanism.
109                      if (hadImage) {
110                          cts.writeTile(address.row, address.column, address.
                                 level,
111                                         tileImage);
112
113                      }
114                  }
115              }
116          }
117          cts.close();
118      }
119
120      public static Rectangle convertCoordinates(BoundingBox imageBounds,
             BoundingBox subImageBounds, int imageWidth, int imageHeight) {
121
122          int x = (int) Math.round((subImageBounds.minx − imageBounds.minx) / (
                 imageBounds.maxx − imageBounds.minx) * imageWidth);
123          int y = imageHeight − (int) Math.round((subImageBounds.maxy −
                 imageBounds.miny) / (imageBounds.maxy − imageBounds.miny) *
                 imageHeight);
124          int width = (int) Math.round((subImageBounds.maxx − subImageBounds.minx
                 ) / (imageBounds.maxx − imageBounds.minx) * imageWidth);
125          int height = (int) Math.round((subImageBounds.maxy − subImageBounds.
                 miny) / (imageBounds.maxy − imageBounds.miny) * imageHeight);
126          Rectangle r = new Rectangle(x, y, width, height);
127          return r;
128      }
129
130      public static void drawImageToImage(BufferedImage source, BoundingBox
             source_bb,
131                                           BufferedImage target, BoundingBox
                                                 target_bb) {
132          double xd = target_bb.maxx − target_bb.minx;
133          double yd = target_bb.maxy − target_bb.miny;
134          double wd = (double) target.getWidth();
135          double hd = (double) target.getHeight();
136          double targdpx = xd / wd;
137          double targdpy = yd / hd;
138          double srcdpx = (source_bb.maxx − source_bb.minx) / source.getWidth();
139          double srcdpy = (source_bb.maxy − source_bb.miny) / source.getHeight();
```

```
140         int tx = (int) Math.round(((source_bb.minx - target_bb.minx) / targdpx)
                );
141         int ty = target.getHeight() - (int) Math.round(((source_bb.maxy -
                target_bb.miny) / yd) * hd);
142
143         int tw = (int) Math.ceil(((srcdpx / targdpx) * source.getWidth()));
144         int th = (int) Math.ceil(((srcdpy / targdpy) * source.getHeight()));
145         Graphics2D target_graphics = (Graphics2D) target.getGraphics();
146
147         //use one of these three statements to set the interpolation method to
                be used
148         target_graphics.setRenderingHint(RenderingHints.KEY_INTERPOLATION,
                RenderingHints.VALUE_INTERPOLATION_BILINEAR);
149
150         target_graphics.drawImage(source, tx, ty, tw, th, null);
151     }
152 }
153
154
155 public class PullTilerThread extends Thread {
156
157     private TileRangeIterator tri;
158     private CachedClusteredTileStream cts;
159     private RawImageReader imageReader;
160
161     public PullTilerThread(TileRangeIterator tri, CachedClusteredTileStream cts
            , RawImageReader imageReader) {
162         this.tri = tri;
163         this.cts = cts;
164         this.imageReader = imageReader;
165     }
166
167     public void run() {
168         while (this.tri.hasMoreTiles()) {
169             TileAddress address = this.tri.getNextTileID();
170             //Compute the geographic bounds of the specific tile.
171             BoundingBox tileBounds = address.getBoundingBox();
172             //get the bounds of the sub-image
173             Rectangle rect = PullTileCreation.convertCoordinates(
174             imageReader.imageBounds, tileBounds, imageReader.imageWidth,
                    imageReader.imageHeight);
175             //extract the image data from the source image
176             BufferedImage subImage = imageReader.getSubImage(rect.x, rect.y,
                    rect.width, rect.height);
177             //create a new empty image
178             BufferedImage tileImage = new BufferedImage(PullTileCreation.
                    TILE_SIZE, PullTileCreation.TILE_SIZE, BufferedImage.
                    TYPE_INT_RGB);
179             //scale the source image to the new image
180             PullTileCreation.drawImageToImage(subImage, tileBounds, tileImage,
                    tileBounds);
181             if (tileImage != null) {
182                 //write the image to the cache
183                 cts.writeTile(address.row, address.column, address.level,
                        tileImage);
184
185             }
186         }
187     }
188 }
```

Listing 12.5 Push-based tile creation for Blue Marble.

```
 1  public class PushTileCreation {
 2
 3      static int TILE_SIZE = 512;
 4
 5      public static void main(String[] args) {
 6
 7          int baseLevel = 8;
 8          String folder = "folder";
 9
10          //create source image records
11          SourceImage a1 = new SourceImage("A1.jpg", new BoundingBox(-180, 0,
                    -90, 90), 21600, 21600);
12          SourceImage b1 = new SourceImage("B1.jpg", new BoundingBox(-90, 0, 0,
                    90), 21600, 21600);
13          SourceImage c1 = new SourceImage("C1.jpg", new BoundingBox(0, 0, 90,
                    90), 21600, 21600);
14          SourceImage d1 = new SourceImage("D1.jpg", new BoundingBox(90, 0, 180,
                    90), 21600, 21600);
15          SourceImage a2 = new SourceImage("A2.jpg", new BoundingBox(-180, -90,
                    -90, 0), 21600, 21600);
16          SourceImage b2 = new SourceImage("B2.jpg", new BoundingBox(-90, -90, 0,
                    0), 21600, 21600);
17          SourceImage c2 = new SourceImage("C2.jpg", new BoundingBox(0, -90, 90,
                    0), 21600, 21600);
18          SourceImage d2 = new SourceImage("D2.jpg", new BoundingBox(90, -90,
                    180, 0), 21600, 21600);
19
20          SourceImage[] images = new SourceImage[] {
21              a1,
22              b1,
23              c1,
24              d1,
25              a2,
26              b2,
27              c2,
28              d2
29          };
30
31          //create output stream to store tiles
32          CachedClusteredTileStream cts = new CachedClusteredTileStream("folder2"
                    , "bluemarble", baseLevel, baseLevel + 1);
33
34          //build base level
35          for (int i = 0; i < images.length; i++) {
36              SourceImage currentImage = images[i];
37              BoundingBox currentBounds = currentImage.bb;
38              //determine the tile bounds specific to each source image
39              long mincol = (long) Math.floor((currentBounds.minx + 180.0) /
                    (360.0 / Math.pow(2.0, (double) baseLevel)));
40              long maxcol = (long) Math.floor((currentBounds.maxx + 180.0) /
                    (360.0 / Math.pow(2.0, (double) baseLevel)));
41              long minrow = (long) Math.floor((currentBounds.miny + 90.0) /
                    (180.0 / Math.pow(
42          2.0, (double) baseLevel - 1)));
43              long maxrow = (long) Math.floor((currentBounds.maxy + 90.0) /
                    (180.0 / Math.pow(
44          2.0, (double) baseLevel - 1)));
45
46              //if the image bounds go beyond the allowed tile bounds, set them
                    to the proper range
47              if (maxrow >= TileStandards.zoomRows[baseLevel]) {
48                  maxrow = TileStandards.zoomRows[baseLevel] - 1;
49              }
50              if (maxcol >= TileStandards.zoomColumns[baseLevel]) {
51                  maxcol = TileStandards.zoomColumns[baseLevel] - 1;
52              }
```

```
53
54              // read the source image from disk
55              BufferedImage bi = null;
56              try {
57                  bi = ImageIO.read(new File(folder + "/" + currentImage.name));
58              } catch (IOException e) {
59                  e.printStackTrace();
60              }
61              // iterate over the current tile bounds and create the tiled images
62              for (long c = mincol; c <= maxcol; c++) {
63                  for (long r = minrow; r <= maxrow; r++) {
64                      TileAddress address = new TileAddress(r, c, baseLevel);
65                      BoundingBox tileBounds = address.getBoundingBox();
66                      // check the cache for a pre-existing tiled image,
67                      BufferedImage tileImage = cts.readTile(address.row, address
                            .column, address.level);
68                      if (tileImage == null) {
69                          // the image wasn't in the cache, so create a new one
70                          tileImage = new BufferedImage(TILE_SIZE, TILE_SIZE,
                                BufferedImage.TYPE_INT_ARGB);
71                          cts.writeTile(address.row, address.column, address.
                                level, tileImage);
72                      }
73
74                      drawImageToImage(bi, currentBounds, tileImage, tileBounds);
75
76                  }
77              }
78          }
79          // iterate over the remaining levels
80          for (int level = baseLevel − 1; level >= 1; level−−) {
81              long curMinCol = 0;
82              long curMaxCol = TileStandards.zoomColumns[level] − 1;
83              long curMinRow = 0;
84              long curMaxRow = TileStandards.zoomRows[level] − 1;
85              // Iterate over the tile set coordinates.
86              for (long c = curMinCol; c <= curMaxCol; c++) {
87                  for (long r = curMinRow; r <= curMaxRow; r++) {
88                      // For each tile, do the following:
89                      TileAddress address = new TileAddress(r, c, level);
90                      // Determine the FOUR tiles from the higher level that
                            contribute to the current tile.
91                      TileAddress tile00 = new TileAddress(r * 2, c * 2, level +
                            1);
92                      TileAddress tile01 = new TileAddress(r * 2, c * 2 + 1,
                            level + 1);
93                      TileAddress tile10 = new TileAddress(r * 2 + 1, c * 2,
                            level + 1);
94                      TileAddress tile11 = new TileAddress(r * 2 + 1, c * 2 + 1,
                            level + 1);
95                      // Retrieve the four tile images, or as many as exist.
96                      BufferedImage image00 = cts.readTile(tile00.row, tile00.
                            column, tile00.level);
97                      BufferedImage image01 = cts.readTile(tile01.row, tile01.
                            column, tile01.level);;
98                      BufferedImage image10 = cts.readTile(tile10.row, tile10.
                            column, tile10.level);
99                      BufferedImage image11 = cts.readTile(tile11.row, tile11.
                            column, tile11.level);
100                     // Combine the four tile images into a single, leveld−down
                            image.
101                     BufferedImage tileImage = new BufferedImage(
102                     TILE_SIZE, TILE_SIZE, BufferedImage.TYPE_INT_RGB);
103                     Graphics2D g = (Graphics2D) tileImage.getGraphics();
104                     g.setRenderingHint(RenderingHints.KEY_INTERPOLATION,
                            RenderingHints.VALUE_INTERPOLATION_BILINEAR);
105                     boolean hadImage = false;
```

```
106         if ((image00 != null)) {
107             g.drawImage(image00, 0, Constants.TILE_SIZE_HALF,
                    Constants.TILE_SIZE_HALF, Constants.TILE_SIZE, 0,
                    0, Constants.TILE_SIZE, Constants.TILE_SIZE, null)
                    ;
108             hadImage = true;
109         }
110         if ((image01 != null)) {
111             g.drawImage(image01, Constants.TILE_SIZE_HALF,
                    Constants.TILE_SIZE_HALF, Constants.TILE_SIZE,
                    Constants.TILE_SIZE, 0, 0, Constants.TILE_SIZE,
                    Constants.TILE_SIZE, null);
112             hadImage = true;
113         }
114         if ((image10 != null)) {
115             g.drawImage(image10, 0, 0, Constants.TILE_SIZE_HALF,
                    Constants.TILE_SIZE_HALF, 0, 0, Constants.
                    TILE_SIZE, Constants.TILE_SIZE, null);
116             hadImage = true;
117         }
118         if ((image11 != null)) {
119             g.drawImage(image11, Constants.TILE_SIZE_HALF, 0,
                    Constants.TILE_SIZE, Constants.TILE_SIZE_HALF, 0,
                    0, Constants.TILE_SIZE, Constants.TILE_SIZE, null)
                    ;
120             hadImage = true;
121         }
122         // save the completed tiled image to the tile storage
                mechanism.
123         if (hadImage) {
124             cts.writeTile(address.row, address.column, address.
                    level, tileImage);
125         }
126     }
127   }
128 }
129 cts.close();
130 }
```

Chapter 13
Case Study: Supporting Multiple Tile Clients

Chapter 9 presented techniques for serving tiled images according to our simple tile protocol and scheme. However, most mapping tools do not support this protocol by default. The purpose of this chapter is to present specialized techniques for tile serving that will support a wide variety of mapping tools. We will build an interface for Google Earth using the Keyhole Markup Language (KML). We will also build an Open Geospatial Consortium (OGC) Web Map Service (WMS) server.

13.1 KML Server

The KML language is a fairly expressive display control language for 3-D globe-oriented mapping tools like Google Earth. It allows for both dynamic positioning of the globe view and dynamic placement of mapping objects on the globe. Mapping objects can include vector features or image overlays. We will be using its image overlay capability to display tiled images on Google Earth.

13.1.1 Static KML Example

Our first example will be a statically generated KML file that links to tiled images from the local computer. The KML file is generated ahead of time and its contents do not change. The KML file in Listing 13.1 creates GroundOverlay objects corresponding to tiles from our Blue Marble tile set, zoom level 2.

Because we have provided the bounding boxes for each tiled image, Google Earth can draw the images on the globe in the correct position. Figure 13.1 shows a screenshot of the static KML file loaded into Google Earth. The KML file requires the image files be stored with it in the same folder. By replacing the file names in the above KML document with URLs we can force Google Earth to retrieve the images not from local disk but directly from a server. We can form URLs using the

J.T. Sample and E. Ioup, *Tile-Based Geospatial Information Systems:*
Principles and Practices, DOI 10.1007/978-1-4419-7631-4_13,
© Springer Science+Business Media, LLC 2010

method presented Section 9.2. Because KML is written in XML, we have to write
the ampersand character & as & in our URL. Instead of:

```
1 <Icon>
2     <href>2-0-2.jpg</href>
3 </Icon>
```

we would put:

```
1 <Icon>
2     <href>http://www.sometileserver.com/tiles?REQUEST=GETTILE&amp;LAYER=
           BlueMarble&amp;LEVEL=2&amp;ROW=0&amp;COLUMN=2</href>
3 </Icon>
```

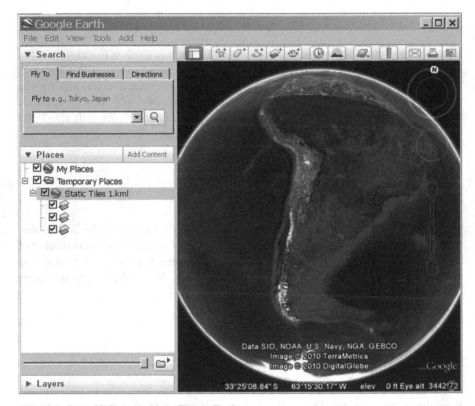

Fig. 13.1 Static KML loaded into Google Earth

This example is effective for loading images into Google Earth, but as the user
zooms in or moves the globe around, the tiled images will not change. It would be
better to create a method that could dynamically load and position tiled images as
the user moves the globe.

13.1.2 Dynamic KML Example

Fortunately, Google Earth allows dynamic content generation based on communication between the client and the server. Using the `<NetworkLink>` KML tag, we can tell Google Earth to load tiled images based on the current globe position, and we can also tell Google Earth to refresh the content each time the globe moves. The following KML snippet creates a NetworkLink object named "bluemarble" and points it to the URL: `http://www.sometileserver.com/tiles/kml`. It tells Google Earth to refresh the network link each time the globe is moved and stops moving.

```
1  <?xml version="1.0" encoding="UTF-8"?>
2  <kml xmlns="http://earth.google.com/kml/2.1">
3      <NetworkLink>
4          <name>bluemarble</name>
5          <Link id="ID">
6              <href>http://www.sometileserver.com/tiles/kml</href>
7              <refreshMode>onChange</refreshMode>
8              <refreshInterval>0</refreshInterval>
9              <viewRefreshMode>onStop</viewRefreshMode>
10             <viewRefreshTime>0</viewRefreshTime>
11             <viewBoundScale>1</viewBoundScale>
12             <viewFormat>BBOX=[bboxWest],[bboxSouth],[bboxEast],[bboxNorth]&amp;
                   HEIGHT=[horizPixels]&amp;WIDTH=[vertPixels]</viewFormat>
13         </Link>
14     </NetworkLink>
15  </kml>
```

The values `bboxWest`, `bboxSouth`, `bboxEast`, `bboxNorth`, `horizPixels`, and `vertPixels` are built-in Google Earth parameters. The `<viewFormat>` tag tells Google Earth to append these values to the base URL for the NetworkLink. When activated, it will take the base URL `http://www.sometileserver.com/tiles/kml` and add parameters in the following manner:

`http://www.sometileserver.com/tiles/kml?BBOX=-176.5, 27.57,-42.23,49.73&HEIGHT=1315&WIDTH=1154`

We can then use these parameters to generate a dynamic version of the KML file presented in the previous section. The code in Listing 13.2 is a simple Java servlet that will dynamically generate a KML document based on the view parameters provided.

13.2 WMS Server

While KML is very effective for integrating mapping content with Google Earth, we want a more general solution to provide tile content to a larger set of mapping clients. The Open Geospatial Consortium (OGC) publishes a set of specifications for standardized communication between clients and servers. The Web Map Service

(WMS) standard provides for simple HTTP based access to imagery and rendered map images [1].

WMS is a simple protocol but is exceptionally powerful. It is supported by many mapping applications include Gaia by The Carbon Project, Google Earth, uDig, and OpenLayers. In general, WMS sessions are a two step process. The first step issues a "GetCapabilities" request to get a list of available layers from a WMS server. The second (and subsequent steps) issues a "GetMap" request to retrieve a map image. GetCapabilities requests are executed by forming a URL in the following manner:

```
http://www.sometileserver.com/wms?REQUEST=
GetCapabilities&VERSION=1.1.1&SERVICE=WMS
```

This request will return an XML document as shown in Listing 13.3. This document provides information about the service provider, methods of access and a list of available layers. Each available map layer is given with information about its map bounds and geospatial projection. Client applications can request map images by forming URLs in the following manner:

```
http://www.sometileserver.com/wms?REQUEST=
GetMap&SERVICE=WMS&VERSION=1.1.1&LAYERS=
bluemarble&FORMAT=image/png&SRS=EPSG:4326&BBOX=-130.
04,-9.20,-52.16,54.75&WIDTH=1225&HEIGHT=1006
```

For our Blue Marble imagery, the URL above should return the image in Figure 13.2.

13.2.1 WMS Servlet Implementation

To create a basic WMS implementation, we need to support only the GetCapabilites and GetMap requests. For our example, we can return a simple XML document for the GetCapabilities request. Implementing the GetMap request will require a little more effort. From the GetMap request URL, we can see the following parameters:

```
BBOX=-130.04,-9.20,-52.16,54.75
```

```
WIDTH=1225
```

```
HEIGHT=1006
```

These parameters illustrate how WMS differs from our previous tiled mapping examples. Our tiled images are created to match a sequence of fixed, discrete scales. In contrast, WMS requests allow users to select maps of varied sizes and resolutions that allow viewing of continuous scales.

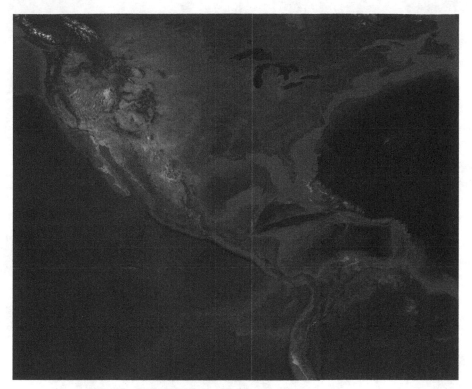

Fig. 13.2 Bluemarbe image resulting from a WMS request.

To use our tile-based data sets to provide a proper GetMap response to this query, we must do two things. First, we need to take the provided bounding box and image dimensions and determine which tiles should be used to create the output image. Second, we have to merge the tiled images into a single composite image. This process is very similar to the one described in Section 3.1.2.

First, we select the scale that we will use. This is done by computing the degrees per pixel for the requested image and finding the closest tile zoom level (Equation (3.3)). Recall that for our tile scheme, each zoom level has its own uniform DPP value. Next, we compute the tiles needed to generate the map (Equation (2.8)). Given the list of tiled images, we can retrieve those images and draw them together in the same image. Equation (2.3) shows how to compute the geographic bounds for each, and Listing 4.10 shows how to draw the tiled images into a single image.

The code in Listing 13.4 shows a simple Java Servlet implementation of a Web Map Service that draws its imagery from a store of tiled images. The example code is intended to support version 1.1.1 of the WMS specification. A production WMS server should implement all versions of the WMS specification. Our example does not validate the query parameters or provide error handling. It also uses a completely static "Capabilities" document; a production version would dynamically generate the Capabilities document to reflect the names and bounds of available layers, which might change.

Another feature of the WMS specification is the "GetFeatureInfo" operation. This allows WMS clients to query map layers for information at a point on the map. GetFeatureInfo requests pass the original coordinates and size of the requested map to the server and the x and y coordinates representing a mouse click. Responses to the GetFeatureInfo query can come in many forms, but will usually provide some text information on map content at the location queried. Developers of tile-based systems might use the GetFeatureInfo operation to provide users with the names of the source files used to make the tile at the location queried.

Listing 13.1 KML file for static inclusion of tiled images.

```
 1
 2  <?xml version="1.0" encoding="UTF-8"?>
 3  <kml xmlns="http://earth.google.com/kml/2.1">
 4      <Document>
 5          <GroundOverlay>
 6              <drawOrder>21</drawOrder>
 7              <Icon>
 8                  <href>2-0-0.jpg</href>
 9              </Icon>
10              <LatLonBox>
11                  <north>0.0</north>
12                  <south>-90.0</south>
13                  <east>-90.0</east>
14                  <west>-180.0</west>
15              </LatLonBox>
16          </GroundOverlay>
17          <GroundOverlay>
18              <drawOrder>21</drawOrder>
19              <Icon>
20                  <href>2-0-1.jpg</href>
21              </Icon>
22              <LatLonBox>
23                  <north>0.0</north>
24                  <south>-90.0</south>
25                  <east>0.0</east>
26                  <west>-90.0</west>
27              </LatLonBox>
28          </GroundOverlay>
29          <GroundOverlay>
30              <drawOrder>21</drawOrder>
31              <Icon>
32                  <href>2-0-2.jpg</href>
33              </Icon>
34              <LatLonBox>
35                  <north>0.0</north>
36                  <south>-90.0</south>
37                  <east>90.0</east>
38                  <west>0.0</west>
39              </LatLonBox>
40          </GroundOverlay>
41
42      </Document>
43  </kml>
```

Listing 13.2 Servlet for dynamic XML content.

```
1   public class KMLServlet extends HttpServlet {
2
3       DataStore dataStore;
4
5       public void doGet(HttpServletRequest request, HttpServletResponse response)
                {
6           String layer = "bluemarble";
7           //set the content type for KML
8           response.setContentType("application/keyhole");
9
10          //get the view format parameters from the query
11          int width = Integer.parseInt(request.getParameter("WIDTH"));
12          int height = Integer.parseInt(request.getParameter("HEIGHT"));
13          String bbox = request.getParameter("BBOX");
14          String[] bboxdata = bbox.split(",");
15          double minx = Double.parseDouble(bboxdata[0]);
16          double maxx = Double.parseDouble(bboxdata[2]);
17          double miny = Double.parseDouble(bboxdata[1]);
18          double maxy = Double.parseDouble(bboxdata[3]);
19
20          //this section determines which zoom level to use
21          double currentDPP = 0.8 * (maxx - minx) / width;
22          double[] zoomLevels = TileStandards.zoomLevels;
23          int scale = 18;
24          for (int i = 2; i < zoomLevels.length; i++) {
25              if (zoomLevels[i] < currentDPP) {
26                  scale = i;
27                  break;
28              }
29          }
30
31          int minscale = dataStore.getMinScale(layer);
32          int maxscale = dataStore.getMaxScale(layer);
33
34          if (scale > maxscale) {
35              scale = maxscale;
36          }
37          if (scale < minscale) {
38              scale = minscale;
39          }
40          //determine the range of tiles to use in the response
41          long startRow = (long) Math.floor(TileStandards.getRowForCoord(miny,
                scale));
42          long endRow = (long) Math.floor(TileStandards.getRowForCoord(maxy,
                scale));
43          long startCol = (long) Math.floor(TileStandards.getColForCoord(minx,
                scale));
44          long endCol = (long) Math.floor(TileStandards.getColForCoord(maxx,
                scale));
45
46          //get the output writer
47          PrintWriter pw=null;
48          try {
49              pw = response.getWriter();
50          } catch (IOException e) {
51              e.printStackTrace();
52          }
53
54          //print document header
55          pw.println("<?xml version=\"1.0\" encoding=\"UTF-8\"?>"
56                  + "<kml xmlns=\"http://earth.google.com/kml/2.1\">"
57                  + "<Document>");
58
59          double degsPerTile = getDegreesPerTile(scale);
60          for (long i = startRow; i <= endRow; i++) {
61              for (long j = startCol; j <= endCol; j++) {
```

```
62                    double tminx = j * degsPerTile - 180.0;
63                    double tmaxx = (1 + j) * degsPerTile - 180.0;
64                    double tminy = i * degsPerTile - 90.0;
65                    double tmaxy = (1 + i) * degsPerTile - 90.0;
66                    pw
67                        . println ("<GroundOverlay>"
68                            + "<drawOrder>21</drawOrder>"
69                            + "<Icon>"
70                            + " <href>http ://www. sometileserver .com/ tiles&amp;
                                       REQUEST=GETTILE&amp; "
71                            + "layer=bluemarble&amp;" + "row=" + i + "&amp; col="
                                       + j
72                            + "&amp; level=" + scale + "</href>"
73                            + "</Icon>"
74                            + "<LatLonBox>"
75                            + " <north>"+ tmaxy + "</north>"
76                            + "<south>" + tminy + "</south>"
77                            + "<east>"  + tmaxx + "</east>"
78                            + "<west>"  + tminx + "</west>"
79                            + "</LatLonBox>"
80                            + "</GroundOverlay>");
81                }
82            }
83            // print document footer
84            pw. println ("</Document></kml>");
85
86        }
87
88        private double getDegreesPerTile (int scale) {
89            double degs = 360.0 / (Math .pow(2, scale));
90            return degs;
91        }
92 }
```

Listing 13.3 Example WMS Capabilities document.

```
1  <WMT_MS_Capabilities version="1.1.1" updateSequence ="0">
2      <Service>
3          <Name>WMS</Name>
4          <Title>Tile_Server</Title>
5          <Abstract>
6          </Abstract>
7          <KeywordList>
8              <Keyword>Some Keywords</Keyword>
9          </KeywordList>
10         <OnlineResource xmlns:xlink="http://www.w3.org/1999/xlink" xlink:type="
                simple" xlink:href="http://www.sometileserver.com/" />
11         <ContactInformation>
12             <ContactPersonPrimary>
13                 <ContactPerson>John Q. Developer</ContactPerson>
14                 <ContactOrganization>Some Company</ContactOrganization>
15             </ContactPersonPrimary>
16             <ContactPosition />
17             <ContactAddress>
18                 <AddressType>postal</AddressType>
19                 <Address>123 Easy Street</Address>
20                 <City>Sometown</City>
21                 <StateOrProvince>BG</StateOrProvince>
22                 <PostCode>31111</PostCode>
23                 <Country>USA</Country>
24             </ContactAddress>
25             <ContactVoiceTelephone>+1 321 456-4200</ContactVoiceTelephone>
26             <ContactFacsimileTelephone>+1 321 456-4443</
                    ContactFacsimileTelephone>
27             <ContactElectronicMailAddress>developer@somecompany.com</
                    ContactElectronicMailAddress>
28         </ContactInformation>
```

```
29          <Fees>none</Fees>
30          <AccessConstraints>none</AccessConstraints>
31      </Service>
32      <Capability>
33          <Request>
34              <GetCapabilities>
35                  <Format>application/vnd.ogc.wms_xml</Format>
36                  <DCPType>
37                      <HTTP>
38                          <Get>
39                              <OnlineResource xmlns:xlink="http://www.w3.org
                                   /1999/xlink" xlink:type="simple" xlink:href="
                                   http://www.sometileserver.com/wms?" />
40                          </Get>
41                      </HTTP>
42                  </DCPType>
43              </GetCapabilities>
44              <GetMap>
45                  <Format>image/png</Format>
46                  <Format>image/jpeg</Format>
47                  <DCPType>
48                      <HTTP>
49                          <Get>
50                              <OnlineResource xmlns:xlink="http://www.w3.org
                                   /1999/xlink" xlink:type="simple" xlink:href="
                                   http://www.sometileserver.com/wms?" />
51                          </Get>
52                      </HTTP>
53                  </DCPType>
54              </GetMap>
55          </Request>
56          <Exception>
57              <Format>application/vnd.ogc.se_xml</Format>
58          </Exception>
59          <VendorSpecificCapabilities />
60          <Layer>
61              <Title>Tile_Server</Title>
62              <Layer queryable="1">
63                  <Name>bluemarble</Name>
64                  <Title>bluemarble</Title>
65                  <SRS>EPSG:4326</SRS>
66                  <LatLonBoundingBox SRS="EPSG:4326" minx="-180.0" miny="-90 0"
                       maxx="180.0" maxy="90.0" />
67                  <BoundingBox SRS="EPSG:4326" minx="-180.0" miny="-90.0" maxx="
                       180.0" maxy="90.0" />
68              </Layer>
69          </Layer>
70      </Capability>
71  </WMT_MS_Capabilities>
```

Listing 13.4 Java Servlet implementation of simple WMS server.

```
1  public class WMSTileServlet extends HttpServlet {
2
3      DataStore tileStorage;
4
5      public void doGet(HttpServletRequest request, HttpServletResponse response)
             {
6          // collect the parameters from the URL
7          String service = request.getParameter("SERVICE");
8          // service should equal WMS
9          String version = request.getParameter("VERSION");
10         // version should equal 1.1.1
11
12         String requestType = request.getParameter("REQUEST");
13         if (requestType.equalsIgnoreCase("GetCapabilities")) {
14             printCapabilities(request, response);
15         }
16
17         if (requestType.equalsIgnoreCase("GETMAP")) {
18             String layers = request.getParameter("LAYERS");
19             // layers should equal blue marble
20             String srs = request.getParameter("SRS");
21             // should be equal to EPSG:4326
22
23             String widthStr = request.getParameter("WIDTH");
24             String heightStr = request.getParameter("HEIGHT");
25             int width = Integer.parseInt(widthStr);
26             int height = Integer.parseInt(heightStr);
27             String format = request.getParameter("FORMAT");
28             // get bounding box from request
29             String bbox = request.getParameter("BBOX");
30             String bbdata[] = bbox.split(",");
31             double minx = Double.parseDouble(bbdata[0]);
32             double miny = Double.parseDouble(bbdata[1]);
33             double maxx = Double.parseDouble(bbdata[2]);
34             double maxy = Double.parseDouble(bbdata[3]);
35             BoundingBox imageBounds = new BoundingBox(minx, miny, maxx, maxy);
36             // compute scale to use
37             double dpp = ((maxx - minx) / width + (maxy - miny) / height) /
                   2.0;
38             double[] standardScales = TileStandards.zoomLevels;
39             int maxscale = tileStorage.getMaxScale("bluemarble");
40
41             int scaleToUse = maxscale;
42             for (int i = 0; i < maxscale; i++) {
43                 if (standardScales[i] < dpp) {
44                     scaleToUse = i;
45                     break;
46                 }
47             }
48             double tileSize = 360.0 / (Math.pow(2, scaleToUse));
49             // calculate bounds of image in tile coordinates
50             long mincol = (long) Math.max(0, Math.floor((minx + 180.0) /
                   tileSize));
51             long maxcol = (long) Math.floor((maxx + 180.0) / tileSize);
52             long minrow = (long) Math.max(0, Math.floor((miny + 90.0) /
                   tileSize));
53             long maxrow = (long) Math.floor((maxy + 90.0) / tileSize);
54
55             // iterate over tile range, retrieve each tile and draw to new image
56             BufferedImage bi = new BufferedImage(width, height,
57             BufferedImage.TYPE_INT_RGB);
58             for (long r = minrow; r <= maxrow; r++) {
59                 for (long c = mincol; c <= maxcol; c++) {
60                     // have to check to see if the cluster actually has any data
                       ...
61                     TileAddress tileAddress = new TileAddress(r, c, scaleToUse)
                         ;
62
63                     BoundingBox tileBounds = tileAddress.getBoundingBox();
64                     byte[] imageData = tileStorage.getTileImage("bluemarble",
65                     scaleToUse, r, c);
66
```

```
67                         if (imageData != null) {
68                             try {
69                                 BufferedImage tileImage = ImageIO.read(new
                                       ByteArrayInputStream(imageData));
70                                 drawImageToImage(tileImage, tileBounds, bi,
                                       imageBounds);
71                             } catch (IOException e) {
72                                 e.printStackTrace();
73                             }
74                         }
75                     }
76                 }
77
78                 //encode image and write to client
79                 try {
80                     ByteArrayOutputStream baos = new ByteArrayOutputStream();
81                     if (format.matches(".*png.*")) {
82                         ImageIO.write(bi, "png", baos);
83                         response.setContentType("image/png");
84                     } else {
85                         ImageIO.write(bi, "jpg", baos);
86                         response.setContentType("image/jpg");
87                     }
88                     ServletOutputStream os = response.getOutputStream();
89                     os.write(baos.toByteArray());
90                     os.close();
91                 } catch (IOException e) {
92                     e.printStackTrace();
93                 }
94             }
95     }
96
97     public static void drawImageToImage(BufferedImage source, BoundingBox
            source_bb, BufferedImage target, BoundingBox target_bb) {
98         double xd = target_bb.maxx - target_bb.minx;
99         double yd = target_bb.maxy - target_bb.miny;
100        double wd = (double) target.getWidth();
101        double hd = (double) target.getHeight();
102        double targdpx = xd / wd;
103        double targdpy = yd / hd;
104        double srcdpx = (source_bb.maxx - source_bb.minx) / source.getWidth();
105        double srcdpy = (source_bb.maxy - source_bb.miny) / source.getHeight();
106        int tx = (int) Math.round(((source_bb.minx - target_bb.minx) / targdpx)
                );
107        int ty = target.getHeight()
108            - (int) Math.round(((source_bb.maxy - target_bb.miny) / yd) * hd)
109            - 1;
110        int tw = (int) Math.ceil(((srcdpx / targdpx) * source.getWidth()));
111        int th = (int) Math.ceil(((srcdpy / targdpy) * source.getHeight()));
112        Graphics2D target_graphics = (Graphics2D) target.getGraphics();
113
114        target_graphics
115            .setRenderingHint(RenderingHints.KEY_INTERPOLATION,
116            RenderingHints.VALUE_INTERPOLATION_BILINEAR);
117
118        target_graphics.drawImage(source, tx, ty, tw, th, null);
119    }
120
121    private void printCapabilities(HttpServletRequest request,
            HttpServletResponse response) {
122        try {
123            response.setContentType("text/xml");
124            PrintWriter pw = response.getWriter();
125            pw.print(capabilitiesContents);
126            pw.close();
127        } catch (IOException e) {
128            e.printStackTrace();
```

```
129          }
130      }
131
132      public static String capabilitiesFile = "capabilities.xml";
133      public static String capabilitiesContents;
134      //read the capabilities file into memory and hold it
135      static {
136          StringBuffer sb = new StringBuffer();
137          try {
138              RandomAccessFile raf = new RandomAccessFile(capabilitiesFile, "r");
139              byte[] data;
140              data = new byte[(int) raf.length()];
141              raf.readFully(data);
142              raf.close();
143              capabilitiesContents = new String(data);
144          } catch (IOException e) {
145              e.printStackTrace();
146          }
147      }
148 }
```

References

1. de La Beaujardière, J.: Web Map Service Implementation Specification. Open Geospatial Consortium Specification pp. 06–042 (2006)

Index